Major Transitions in Evolution

Anthony Martin, Ph.D.
John Hawks, Ph.D.

THE
GREAT
COURSES™

PUBLISHED BY:

THE GREAT COURSES
Corporate Headquarters
4840 Westfields Boulevard, Suite 500
Chantilly, Virginia 20151-2299
Phone: 1-800-832-2412
Fax: 703-378-3819
www.thegreatcourses.com

Anthony Martin, Ph.D.

Professor of Practice, Department of
Environmental Studies, Emory University

D r. Anthony (Tony) Martin was born in Terre Haute, Indiana. While growing up there, he developed an early interest in biology, paleontology, and geology, spending much of his time outdoors. His earned his B.S. in Geobiology at St. Joseph's College (Indiana), his M.S. in Geology at Miami University, and his Ph.D. in Geology at the University of Georgia. Dr. Martin has been at Emory University since 1990, where he has taught many classes in geology, paleontology, environmental science, and evolutionary biology. While there, he has won university-wide awards for his outstanding teaching and has often been recognized by student societies for his excellence in teaching. He has a special fondness for teaching students outside the classroom and has taught students in the field in a wide variety of places—such as Georgia, the Bahamas, Texas, Arizona, and Australia—and on subjects ranging from geology to ecology.

Dr. Martin's research interests are mainly in ichnology, which is the study of modern and fossil traces, such as tracks, trails, burrows, nests, and other signs of behavior. Through this research, he can discern how behavior has evolved in various lineages of animals through time, such as in trilobites, insects, crayfish, fish, amphibians, reptiles, and dinosaurs. For example, he was the codiscoverer of the only known burrowing dinosaur (*Oryctodromeus cubicularis*), found the oldest known dinosaur burrows in the geologic record, and named the oldest known fossil crayfish in the Southern Hemisphere (*Palaeoechinastacus australianus*). In recent years, his field research has been mainly in 4 places—the Georgia coast, the Bahamas, Montana, and Australia—though he has done fieldwork in more than 15 countries and more than 30 U.S. states.

Dr. Martin is the author of 3 books: *Introduction to the Study of Dinosaurs* (now in its second edition), an undergraduate general-science textbook on dinosaurs; *Trace Fossils of San Salvador*, a field guide to the trace fossils of the Bahamian island attributed to Columbus's first landfall in the New World; and *El dinosaurio que excavó su madriguera* (The dinosaur that dug its burrow), a public-outreach book about the first known burrowing dinosaur, translated into Spanish. He is currently finishing a fourth book, titled *Life Traces of the Georgia Coast*, an overview of plant and animal traces of the Georgia coastal barrier islands and their applications for interpreting the geologic record.

Dr. Martin has also authored more than 30 peer-reviewed papers and book chapters, nearly 100 abstracts, and several public-outreach articles. He is a much-sought speaker for local venues in Atlanta, as well as at universities, natural history museums, and other public venues. He enjoys discussing paleontology and other sciences with the popular media and is often quoted in science news reports for his opinions on the latest discoveries in paleontology.

In terms of hobbies and other pastimes, Dr. Martin is an avid tracker and has taken several tracking courses from nationally recognized instructors. He also loves travel, hiking, biking, and other activities that get him outside and in natural settings. ■

John Hawks, Ph.D.

Associate Professor of Anthropology
University of Wisconsin–Madison

D r. John Hawks was born and raised in Norton, Kansas. In his youth, he followed science but did not seem destined for a career in it. At Kansas State University, he studied English and French literature, earning a B.A. But a chance to teach bone labs inspired him to follow through with a dual degree in Anthropology. Making the switch to science, he went on to earn his M.S. and Ph.D. in Anthropology at the University of Michigan, specializing in Paleoanthropology.

After a postdoctoral fellowship at the University of Utah specializing in human genetics, Dr. Hawks joined the faculty of anthropology at the University of Wisconsin–Madison. He teaches courses across the whole range of students, from freshmen to graduate students. The subjects range from large undergraduate introductory courses in biological anthropology to more specialized topics in human evolution and anthropological genetics.

Dr. Hawks has been awarded several grants for innovation and use of technology in his teaching. His courses have included online collaborative writing projects and computer-based laboratory exercises in genetics. He has incorporated 3-dimensional imaging technology in his fossil bone labs, allowing introductory students to have access to the best-quality research data. Dr. Hawks has mentored independent research projects for dozens of undergraduates. His graduate students have had notable successes in presenting and publishing research and winning grants to support their work.

Early in his career, Dr. Hawks focused mainly on fossil and archaeological evidence for our evolution. But as the Human Genome Project was completed, he became one of the first paleoanthropologists to use genetic and fossil information together to test hypotheses about human prehistory. The genetic record has begun to yield new information about every period of human evolution, from our initial divergence from other lineages of apes up to the last 10,000 years. Dr. Hawks's research has examined this entire time span; he has published research papers on both the earliest possible human ancestors and very recent evolution in historic times. He is truly unique in the way he combines traditional study of fossil evidence with new approaches from genetics. His ability to draw on both kinds of analysis has led to new insights about our recent evolution.

Dr. Hawks's work on the last segment of our evolutionary history has achieved the most impact. He documented the accelerating pace of selection pressures on humans living after the advent of agriculture and connected the rate of change to the growth of human populations. Along with the modeling and analysis of genetic sequences, this study has included the study of Bronze Age and later skeletal samples from Europe, East Asia, and Africa. His work has taken him around the world to examine skeletal remains of both historic and prehistoric populations.

His work on Neandertals has also broken new ground in paleoanthropology. His work in theoretical genetics showed the substantial likelihood of interbreeding of humans and Neandertals, together with the conditions for recognizing Neandertal genes within contemporary populations. These predictions were later confirmed by the direct sequencing of DNA from ancient Neandertal bones. Together with his students, he is examining the function of Neandertal genes and the ways that human populations evolved across the last 50,000 years. His work in this area was featured in the National Geographic Channel documentary *The Neanderthal Code*.

Dr. Hawks has become well-known for writing one of the top blogs on science, where people can follow his descriptions of the latest science in paleoanthropology. His site is visited more than 7000 times a day, and in any given month, his blog is read by people in more than 150 countries. He travels widely to lecture about human origins, has given hundreds of radio and press interviews on the topic, and is a *Science Saturday* regular for the online interview show Bloggingheads.tv. ■

Table of Contents

Table of Contents

Table of Contents

Major Transitions in Evolution

Scope:

This course is intended as an overview of significant evolutionary transitions in the 4.0 billion-year history of life. The first lecture introduces students to basic concepts of macroevolution, including the factors responsible for significant changes in life over the course of geologic time: geographic isolation, genetic drift, environmental change, and natural selection. One of the best ways to study evolutionary transitions is through paleontology (the study of ancient life and fossils), the topic of Lecture 2. The second lecture also covers major eras and periods of the geologic time scale and explains how radiometric age dating confirms the ages of rocks and fossils.

Lecture 3 discusses the evolution of one-celled organisms, delving into the two major types of cells (prokaryotes and eukaryotes) and their differences. Knowing these distinctions, we explore the likely role of symbiosis in the evolution of prokaryotes into eukaryotes and examine the fossil evidence for one-celled organisms from about 3.5 to 1.5 billion years ago. Lecture 4 reviews the early evolution of multicellular animals (metazoans), the fossils of which are represented worldwide in rocks from about 600 to 550 million years ago. Lecture 5 is about the Cambrian period (545 to 505 million years ago) and the next significant transition in the evolution of animals: the formation of mineralized skeletons. Skeletons were probable evolutionary responses to predation but are also linked to the changing chemistry of the Cambrian ocean and atmosphere. Lecture 6 explores the evolution of some Cambrian invertebrate animals into animals with primitive "backbones" and other anatomical innovations, leading to the first chordates and vertebrates.

In Lecture 7, we review the evolutionary challenges faced by algae, fungi, plants, and animals that adapted to terrestrial environments during the early part of the Paleozoic era (about 500 to 350 million years ago). Connected to this topic is the early evolution of insects, the most successful group of animals today. In Lecture 8, we look at insect origins from the fossil record of the Devonian period (410 to 360 million years ago) and examine evidence that some insects evolved into the first known flying animals during the Carboniferous period (360 to 285 million years ago). The evolution of seed plants in the Devonian period resulted in the first true forests, the subject of Lecture 9. Also during the Devonian period, four-limbed vertebrates (tetrapods) evolved from lobe-finned fish. In Lecture 10, we learn about the fossil evidence for this transition, the anatomy of lobe-finned fish and amphibians, and the reflections of their traits in all tetrapods, including humans. Later, in the Carboniferous period, egg-laying reptiles evolved from amphibians, which allowed tetrapods to live in dry-land environments; this topic is discussed in Lecture 11.

Lecture 12 deals with the early evolution of dinosaurs from ancestors shared by modern crocodilians and flying reptiles (pterosaurs), before they evolved into the largest land-dwelling carnivores and herbivores of all time. In Lecture 13, we study the origins and diversification of a wide variety of reptiles that swam in the seas and flew through the air during the Mesozoic era (250 to 65 million years ago), including mosasaurs, plesiosaurs, icthyosaurs, and pterosaurs. Lecture 14 deals with birds, which evolved from theropod dinosaurs during the Jurassic and Cretaceous periods (about 150 to 65 million years ago). Hence, this lecture emphasizes the evolution of feathers and flight in dinosaurs that proliferated into the diversity of birds seen today. On land during the Cretaceous period (145 to 65 million years ago), a major transition between non-flowering and flowering plants took place that resulted in the coevolution of their pollinators. Lecture 15 takes us back to the first primitive flowering plants, while also considering how pollinating animals (especially insects) must have evolved along with the bearers of fruit. Lecture 16 covers the reptiles of the Permian period (285 to 250 million years ago) that evolved mammalian traits, as well as the fossil record for the arrival of the first true mammals, about 225 million years ago (in the Triassic period). This lecture also offers an overview of the major mammal groups.

The mass extinction that ended the Cretaceous period 65 million years ago opened many niches for surviving species of plants and animals, including mammals. Among the evolutionary transitions that took place about 50 million years ago was that of land-dwelling hoofed mammals to marine environments, which led to the evolution of modern whales. Lecture 17 explores the fossil evidence for this transition, highlighting anatomical features that connect these massive animals with their smaller terrestrial ancestors. Just before this time, many kinds of mammals took to the trees, developing adaptations to newly evolved fruits, flowers, and leaves. These developments resulted in the first primates about 60 million years ago, the topic of Lecture 18. By about 45 million years ago, some of these primates evolved into the ancestors of today's monkeys and apes. Lecture 19 describes these primates, their adaptability, probable social habits, and dispersal to new habitats. Lecture 20 deals with the evolution of some ape lineages from 12 to 7 million years ago, some of which began to walk upright, and describes their different sizes, diets, and lifestyles.

The evolutionary path to humanity is the subject of Lecture 21, in which we review the first stone tools, from 2.6 million years ago, marking a change to a human-like social and cognitive system characterized by hunting and gathering. Soon, the descendants of these toolmakers spread more widely than earlier hominids. Lecture 22 looks at how modern humans, once evolved, began to cross Africa about 100,000 years ago, dispersing into Europe and Asia and challenging Neandertals, which are now described by the Neandertal genome. Lecture 23 covers rapid changes in modern human genes, spurred by our high-density agricultural lifestyle and epidemic diseases, as well as human impacts on the Earth's ecology. With this perspective in mind, Lecture 24 summarizes the common themes of major evolutionary transitions, while providing a few examples of how fossils, artifacts, and other evidence help us to better understand the history of life. ∎

Macroevolution and Major Transitions
Lecture 1

Evolution is both a fact and a theory. We've watched these changes happen; that's the fact part. We have an explanation for how it happens; that's the theory part.

Evolution is a good subject for anyone who is interested in looking at life and the present through the lens of deep time. "Deep time" refers to intervals so vast as to be beyond anything we can directly experience in everyday life. We study deep time principally through the sciences of geology and paleontology.

At a simple level, evolution is change over time. Often, evolution is divided into **microevolution**, which refers to change within a species, and **macroevolution**, when one species changes into another. Macroevolution is also understood more broadly as larger changes that happened over many generations and resulted in major transitions, such as the development of a novel means for adapting to a radically new environment. Consider, for example, plant seeds or enclosed animal eggs as adaptations to air environments.

A list of major evolutionary transitions would include the development of eukaryotic cells, multicelled animals, skeletons, life on land, four-legged vertebrates, insect flight and seed plants, enclosed animal eggs, flowering plants, pollinating insects, mammals, and live birth. There are also several important developments involved in the overall transition from tree-dwelling primates to humans capable of documenting the fossil record and understanding how it relates to evolution.

As an example of a major transition in evolution, let's consider marine lobsters and freshwater crayfish. They are quite similar in appearance, but all species of crayfish live in freshwater environments, while lobsters live only in saltwater. In the late 19th century, Thomas Huxley studied the question of how these two animals diverged, but he didn't have knowledge of genetics, plate tectonics, or even fossils. In the 21st century, molecular biologists proposed that crayfish and lobsters diverged from a common ancestor about 270 million years ago, a time when the continents of the Earth were still united. About 180 million years ago, crayfish split into northern and southern hemisphere groups; plate tectonic reconstructions show that this timeframe correlates with the breakup of the southern continents. In 2008, the fossil of a new species of crayfish was discovered, confirming predictions about the location and timing of their transition from lobsters.

Both microevolution and macroevolution have been observed many times in living populations, in many organisms, and in both natural and laboratory settings. For example, speciations have been observed in many species of plants and insects. In fact, many evolutionary scientists consider macroevolution to be the end result of microevolution.

If we really want to understand macroevolution, we can't rely on human time. We've been around as a species for less than 200,000 years or so. Most of the major transitions in evolution occurred in the pre-human past, and we must go back to the fossil record for that evidence. With fossils, we're

The rise and subsequent fall of the dinosaurs are among the most famous of evolutionary transitions, but these major shifts are only two among many.

not observing the actual processes of macroevolution but the products of major changes and lineages of organisms.

Among the factors that are responsible for the macroevolutionary changes we will discuss in this course are geographic isolation, genetic drift, local or global environmental change, and

© Hemera/Thinkstock.

Crayfish fossils have been used to confirm hypotheses about evolution and plate tectonics.

mass extinctions. The effects of these factors can be summarized by one neat little phrase coined by Charles Darwin: natural selection. Of course, our understanding of how the Earth, its life, and its environments have changed through time has improved immensely since Darwin through such new sciences as plate tectonics, developmental biology, ecology, and genetics. ■

Important Terms

macroeveolution: Change from one species to another; more broadly understood as larger changes that occurred over many generations and resulted in major transitions.

microevolution: Change within a species; more precisely, change in the relative proportions of genes over generations within a species.

Suggested Reading

Levinton, *Genetics, Paleontology, and Macroevolution.*

Prothero, *Evolution.*

1. Definitions aside, are the differences between macroevolution and microevolution all that signficant, or are these simply terms that help us to better understand gradations in evolution?

2. Other sciences since the time of Darwin and Huxley have improved macroevolutionary theory. What are some examples from paleontology, geology, and developmental biology during the past year that contributed further to macroevolutionary theory?

Paleontology and Geologic Time
Lecture 2

To better overcome the challenges of confronting deep time when studying evolution, paleontology and geology are the most important sciences.

A fossil is any evidence of ancient life, meaning 10,000 years old or more. Fossils can be placed into three categories. First, body fossils are the actual bodily remains of organisms and offer direct evidence of ancient life. Next are trace fossils, which show some aspect of behavior and offer indirect evidence of ancient life. The third category encompasses chemical fossils, which are organic compounds signifying that once-living material was present.

Contained within the geologic record is a definite ordering of fossils and rocks that is consistent, predictable, and useful to an understanding of evolution. This ordering of fossils, called biological succession, was the result of extinctions and evolution. Since the time of Darwin, geologists and paleontologists have used a well-tested method for dating rocks worldwide, and much of this dating is based on relative ages of fossils. This technique led to the development of the geologic timescale. The broadest time unit in a geologic timescale is the eon, followed by the era, period, and epoch. The boundaries between these time units are based on worldwide extinctions and evolutions of species.

The eons of geologic time in order from oldest to youngest are **Archean**, **Proterozoic**, and **Phanerozoic**. The Phanerozoic is the eon that contains the most abundant and easily identifiable fossils and encompasses most of the major transitions covered in this course. This eon is subdivided into three eras: **Paleozoic**, **Mesozoic**, and **Cenozoic**, based on mass extinctions recorded in the fossil record. Major evolutionary transitions that occurred during the Paleozoic era include the development of skeletons and invertebrates and vertebrates, the first land plants and animals, the first forests and flying insects, enclosed eggs in reptiles, and the early evolution of the ancestors of mammals.

The Mesozoic (or "middle life") era is the time of dinosaurs and many other large reptiles. Major transitions that occurred during the Mesozoic era include the evolution of the dinosaurs and mammals, the evolutionary radiation of some reptiles to the seas and skies, the origin of birds from dinosaurs, and the coevolution of flowering plants with pollinating insects.

Contained within the geologic record is a definite ordering of fossils and rocks that is consistent, predictable, and useful to an understanding of evolution.

The Cenozoic (or "new life") era has two periods. Major transitions here include the evolution of whales from land-dwelling mammals; the origin and early evolution of primates; the divergence of some apes into a lineage that led to our genus, *Homo*; and the evolution of *Homo* into our species.

Radiometric age dating is a well-established method for dating rocks that is based on the fact that radioactive elements decay at constant rates. Decay of what is called a parent element eventually results in a final stable byproduct, known as the daughter element. Rates of decay are then derived from ratios of parent isotopes to daughter isotopes. When that ratio reaches one-half for each of them, we call that a half-life. Radiometric age dates can be discerned from volcanic ash deposits, lava flows, or other igneous rocks that may be interspersed with sedimentary rocks that bear fossils.

The important thing to remember here is that great spans of time are sometimes needed for major transitions to take place. In fact, some evolutionary processes may have operated under time spans so vast that we have a tough time pinpointing exactly when a major transition took place. ■

Important Terms

Archean: The period of time from 3.8 to 2.5 billion years ago; at least one continent began to form and the most primitive single-celled life, such as bacteria and prokaryotes, evolved during this time.

Cenozoic: The period of time from 65 million years ago to the present.

Mesozoic: The period of time from 251 to 65 million years ago.

Paleozoic: The period of time from 543 to 251 million years ago.

Phanerozoic: The period of time from 543 million years ago to the present.

Proterozoic: The period of time from 2.5 billion to 543 million years ago.

Suggested Reading

Levin, *The Earth through Time*.

Stanley, *Earth System History*.

Questions to Consider

1. Given the types of evidence from the fossil record, which do you think is more important for figuring out macroevolution: chemical fossils, body fossils, or trace fossils? Or is this more of a matter of using "the right tool for the right job"? Think about at least three separate instances where each type of fossil would serve as the best form of evidence.

2. Think about how relative and absolute age dating are combined to aid in our understanding of evolutionary transitions. What if you only had one of these tools at your disposal: could you still discern any major evolutionary transitions? Explain why or why not.

Single-Celled Life—Prokaryotes to Eukaryotes
Lecture 3

What we have observed about life and stable carbon isotopes is that living things tend to concentrate carbon-12 relative to carbon-13. This means that the ratio should differ noticeably from a ratio of these two isotopes in the absence of life. This signature of life shows up distinctly in the ratio between the two isotopes. Certain ratios are only caused by life.

All life known on Earth today is based on the same macromolecular building blocks of genes and proteins. Life begins with prokaryotic cells, which include two groups: **bacteria** and **archaea**. Prokaryotes have the following features: They are small; lack a nucleus or nuclear envelope, which means that the chromosomes are scattered throughout the cell; lack organelles with membranes; are typically expressed as one-celled life; and reproduce asexually. The transition from **prokaryotic** to **eukaryotic cells** makes possible all life other than bacteria and archaea.

Eukaryotic cells have the following traits: They are larger; have a definite nucleus with a nuclear envelope, which means that the chromosomes are enclosed; have complex organelles, such as mitochondria and chloroplasts; and reproduce sexually. Because of their primitive traits and smaller sizes, prokaryotic cells clearly evolved into larger and more complex eukaryotic cells at some time in the geologic past.

Prokaryotes are considered the most primitive examples of life. They are the only organisms we see today that can live in extreme environments with very hot, very cold, or very low-oxygen conditions. Prokaryotes are also very simple compared to eukaryotic cells in their organization of genetic material, as well as their lack of organelles and asexual reproduction.

Stromatolites are trace fossils of prokaryotes. They are layered structures in sedimentary rocks, probably formed by colonies of cyanobacteria. Chemical and body fossils offer additional evidence of prokaryotes. Chemical fossils, also called biomarkers, are elements or compounds in rocks that indicate the former presence of life. For early life, one of these biomarkers is provided by stable carbon isotope ratios. Unfortunately, we don't have a really good way of knowing exactly what types of prokaryotes were responsible for these geochemical signatures.

When do eukaryotic cells appear in the fossil record? Because of plate tectonics, weathering, and other earth processes, Precambrian fossils are the least likely to be preserved. But we know that such fossil evidence should be found in sedimentary rocks that formed in aquatic environments. Estimates for the probable evolutionary divergence between prokaryotes and eukaryotes range from about 3.9 to 2.2 billion years ago. One mode for the evolution of eukaryotes is endosymbiosis, which literally means "living together within." In this instance, the term refers to multiple prokaryotic cells fusing to form eukaryotic cells.

Prokaryotes use asexual means for reproduction, such as binary fission or gene transfer, while eukaryotes begin with mitosis. In this form of cell division, chromosomes containing DNA are duplicated from a parent cell into two daughter cells also having an identical number of chromosomes. Perhaps about 1.5 billion years ago, another form of cell division, meiosis, also evolved in eukaryotic cells. The resulting greater genetic variability in eukaryotes enabled them to better adapt to changing environments.

The evolution of eukaryotes from prokaryotes changed the world—literally. The increase in algal abundance caused an increase in oxygen content of the oceans and atmosphere from photosynthesis. The earliest fungi evolved about 1.5 billion years ago; these new eukaryotes filled a niche as decomposers in marine environments. Later on, they would fill the same role in terrestrial environments. Eukaryotic cells capable of movement increasingly organized into multicellular colonies and even evolved into multicellular organisms— the earliest animals. ∎

Important Terms

archaea: A group of singled-celled prokaryotes.

bacteria: Prokaryotes that affect human health.

eukaryotic cells: Single-celled organism with the following traits: larger in size than prokaryotes; having a definite nucleus with a nuclear envelope, which means that the chromosomes are enclosed; having complex organelles; and reproducing sexually

prokaryotic: One-celled organism with the following features: small in size; lacking a nucleus or nuclear envelope, which means that the chromosomes are scattered throughout the cell; lacking organelles with membranes; and reproducing asexually.

Suggested Reading

Margulis and Dolan, *Early Life*.

Rizzotti, *Early Evolution*.

Questions to Consider

1. Eukaryotic cells have some advantages over prokaryotic cells. These advantages must have led to their natural selection from prokaryotes nearly two billion years ago. Nonetheless, how are some extant prokaryotes still better adapted for their respective environments than eukaryotes?

2. How could further study of modern stromatolites, such as those in Australia, the Bahamas, and Abu Dhabi, give us more insights into their importance in the history of life?

Metazoans—The Earliest Multicellular Animals
Lecture 4

The currently reigning hypothesis for the origin of animals is that the first ones descended from colonial eukaryotes that had flagella, these little whip-like appendages that you see in single cells.

The more complex cells of eukaryotes set the stage for the evolution of multicellular eukaryotes called **metazoans**. The evolutionary transition from single-celled eukaryotes, also known as protozoans, had to have happened during the latter part of the Proterozoic eon, between 1 billion and about 600 million years ago.

The eukaryotic cells of animals are quite different from those of plants. For example, animal cells lack chloroplasts, which means they cannot manufacture their own food using sunlight. Eukaryotic cells of animals also have cell membranes, while plants have stiff cell walls. Thus, animals have more plasticity in their growth history.

One way to study the evolution of organisms is by calculating **molecular phylogenies**. These are evolutionary lineages based on genetic differences in molecules of modern organisms. DNA and RNA contain genes that are inferred to have changed through regularly occurring mutations over the course of time. Some of these changes were so regular, in fact, that they are referred to as molecular clocks. The overarching principle of a molecular clock is that the amount of genetic change, or genetic distance, between two organisms corresponds to the amount of time elapsed since they diverged from each other. Molecular clocks applied to the divergence of animals from single-celled eukaryotes yield dates of about 1 billion years ago.

In estimating the initial evolution of metazoans, an exciting recent development comes from chemical fossils. Compounds called "scarrings" were found in a sedimentary deposit in Canada dating from about 635 million years ago. These document that certain types of sponges lived at that time. Sponges or similar animals were probably the first metazoans.

Metazoans are divided into two main groups based on cellular organization and symmetry. The first group, **Radiata**, is represented by radially symmetrical animals, such as sponges and true jellyfish. The second group, **Bilateria**, is represented by bilaterally symmetrical animals, such as humans. Bilaterians are believed to have evolved from ancestors resembling jellyfish.

Metazoan body and trace fossils are most common in Precambrian deposits from about 580 to 550 million years ago. Body fossils tell us that most early metazoans must have had

Sponges are some of the oldest multicellular eukaryotes, or metazoans.

relatively stiff, tough exteriors, although they lacked shells. Some of these early organisms include *Spriggina*, *Dickinsonia*, *Pteridinium*, *Tribrachidium*, and *Kimberella*. Metazoan trace fossils include shallow horizontal burrows, surface trails, and scrape marks. The ability to move implies that some of these early animals had evolved internal musculature.

All early metazoans (from the Ediacaran period) lived in shallow marine environments, and most forms were sedentary and closely associated with the sediment bottom. A few mobile forms were moving on or just below the sediment surface. They grazed on algal mats or fed on organic material while burrowing horizontally underneath the surface.

What caused the seemingly sudden appearance of abundant metazoans in the fossil record from about 600 to 550 million years ago? It's likely that these metazoans represent a relatively rapid evolution of animals following a long period of global cooling. After this period of glaciation, the onset of widespread warm, shallow seas and the breakup of the supercontinent Rodinia allowed life to thrive and diversify under new selection pressures. ■

Important Terms

Bilateria: One of the two main groups of metazoans; represented by bilaterally symmetrical animals, such as humans.

metazoans: Multicellular eukaryotes.

molecular phylogenies: Evolutionary lineage based on genetic differences in molecules of modern organisms.

Radiata: One of the two main groups of metazoans; represented by radially symmetrical animals, such as sponges and true jellyfish.

Suggested Reading

Fedonkin, *The Rise of Animals*.

Vickers-Rich and Komarower (editors), *The Rise and Fall of the Ediacaran Biota*.

Questions to Consider

1. Some of the Ediacaran fossils have traits different from or similar to modern animals. Is it possible that some of these fossils represent unique "experiments" in evolution, and actually are unrelated to any modern organisms? Think about how you could argue either for or against this idea.

2. Consider the disagreements between molecular clocks and the fossil record for Proterozoic metazoans. What are some explanations for these discrepancies, and how could they be reconciled with further research, particularly through fossils?

The Development of Skeletons
Lecture 5

As we can see from the Burgess Shale fauna, animals had developed spikes and plates for protection, and some had these well-developed eyes for seeing predators. Predators, in turn, also had eyes for seeing prey from afar, and they developed appendages and mouth parts for getting past some defenses.

The development of skeletons or mineralized tissues in invertebrate animals points to some biological factors that might have driven a seemingly accelerated evolution of animals just after the Ediacaran period.

One of the most important fossil assemblages from the Cambrian period is the Burgess Shale fauna. It has the remains of soft-bodied animals from about 505 million years ago, but some of these show evolutionary responses to hard-bodied animals. Recent research has offered new insights into how these soft-bodied animals evolved in the presence of newly evolved animals with minerals in their bodies. For example, the organism *Wiwaxia* has numerous overlapping plates and spikes likely made of hardened organic compounds, which it obviously developed to deter predators.

The process by which organisms take elements out of their surrounding environments and precipitate them as minerals in their bodies is called **biomineralization**. Probably the most common examples of this process are seen in molluscan shells. The mineral calcite is commonly found in some algae, in protozoans, and in invertebrate animals today. Calcite also contributed to the skeletons of organisms early in the Cambrian period.

Biomineralization is most commonly expressed in invertebrate animals as either calcite or aragonite in their exoskeletons or endoskeletons. In vertebrates, biomineralization is usually expressed as apatite in teeth and endoskeletons. Common fossils with mineralized parts from the first 100 million years of the Phanerozoic era include those from molluscans, brachiopods, corals, arthropods, and echinoderms.

The first step in the development of biomineralization may have been a biochemical response to changing ocean chemistry, that is, an oversaturation of the world's oceans with carbon dioxide and calcium toward the end of the Precambrian period. The least energetic way for organisms to process the excess calcium and bicarbonate in seawater was to combine them into one precipitate, calcium carbonate. Similarly, high amounts of phosphorus and phosphates caused organisms to precipitate apatite.

At the same time or soon after, natural selection was accelerated by the evolution of predators. Animals that were favorably selected just happened to have mineralized tissues already in place, not from predation but from adapting to ocean chemistry. Still, it's important to note that biomineralization evolved multiple times and in separate lineages of animals. Thus, biomineralization is an example of convergent evolution, which occurs when different lineages of organisms arrive at the same or similar adaptive solutions to selection pressures. As predation became more common, organisms that could secrete shells or other skeletal protection would have been favorably selected.

Corals are examples of invertebrate biomineralization, a process that emerged in the Phanerozoic era.

This brings us to the Red Queen hypothesis, according to which coevolution occurs when one cohort—say, the prey—evolves in response to selection pressure from a second cohort—the predator. Interestingly, the ordering of the two can be switched, which results in selection pressure on both cohorts.

As a result of biomineralization, reefs changed from those that were composed of just stromatolites, or sponges, to full-fledged coral reefs, establishing the foundations for marine ecosystems that we still see today. By the middle Cambrian period, predators had evolved into these ecosystems and were now a normal part of the seascape. ∎

Important Term

biomineralization: The process by which organisms take elements out of their surrounding environments and precipitate them as minerals in their bodies.

Suggested Reading

Gould, *Wonderful Life.*

Mann, *Biomineralization.*

Morris, *The Crucible of Creation.*

Questions to Consider

1. Predation is often invoked as a major selection pressure that led to invertebrate animals developing skeletal tissues late in the Proterozoic eon and early in the Paleozoic era, about 550–500 million years ago. Did this factor precede or follow the changes in oceanic chemistry necessary for these animals to make skeletons in the first place, and how would you test either hypothesis?

2. Since the Burgess Shale fauna was discovered in Cambrian rocks from British Columbia early in the 20th century, similar fossils in Cambrian rocks from elsewhere in the world have been found. Do these fossils make the Burgess Shale fossil assemblage less or more "special" in an evolutionary sense?

The Rise of Vertebrates
Lecture 6

Based on a combination of what we know from fossils and modern biology, we now have a much better handle on how echinoderms, hemichordates, urochordates, cephalochordates, and craniates—like me—all share a fascinating story of evolution since the Precambrian that certainly became a lot more detailed in the Cambrian.

The fossil Clydagnathus, found in 1983, offers clues to the evolution of vertebrates from invertebrates. This fossil had an assemblage of conodonts (small tooth-like structures) that was directly associated with the impression of a long, thin, eel-like animal that belonged to a group called chordates. This group of animals was very successful in its evolution starting in the Cambrian period and for more than 300 million years afterward.

Chordates have the following important traits: openings or slits at the region of the pharynx (often called pharyngeal gill slits); a structure called a notochord that runs the length of the animal to support its nerve chord; and a dorsally located nerve chord. Vertebrates are chordates, as are some invertebrates.

The "how" for the transition between invertebrates and vertebrates is perhaps easiest to answer by simply examining modern animals that have the same characteristics we would expect to find in fossils representing those transitions. Ideally, an animal that's in between an invertebrate and a vertebrate will have some sort of blend of anatomical traits shared between invertebrate chordates and primitive vertebrates.

Some non-chordate invertebrates happen to share a few anatomical traits with chordates. These hemichordates also share common characteristics with echinoderms, such as star fish and sea urchins. In turn, both of these non-chordates share a common ancestor with chordates. It might be difficult to imagine a sea cucumber or a sand dollar as a distant relative to us, but they are.

Vertebrate chordates are placed within a group called **Craniata**. Closely related invertebrate animals consist of two groups, Urochordata and Cephalochordata. Cephalochordates living today are known as lancelets, or amphioxus, and these are regarded as modern examples of immediate ancestors to the most primitive vertebrates. Lancelets have a streamlined, blade-like body; pharyngeal gill slits; a dorsal notochord; muscles arranged in myomeres, and a tail that extends past the anus; however, they have no mineralized tissues. Craniata include vertebrates, and most modern vertebrates are descended from jawed fish. As the name implies, craniates have heads supported by skulls, but these can be composed of either cartilage or bone. We see several fossils that serve as excellent examples of early chordate and vertebrate evolution. Hicuella is a good candidate for a transitional fossil between non-vertebrate chordates and true vertebrates. The oldest known vertebrate from the fossil record may be Amylecomingia or Hycuekthes.

Some non-chordate invertebrates happen to share a few anatomical traits with chordates.

Molecular clock data suggest that the divergence of chordates from non-chordates and invertebrates from non-vertebrates happened in the Proterozoic eon, between 750 and 800 million years ago. Unfortunately, fossils of the descended lineages either were not preserved or haven't yet been found.

Other evidence also shows evolutionary linkages between invertebrates and vertebrates. For example, the swimming larva of echinoderms and hemichordates both live in marine environments and are strikingly similar in their overall anatomy and forms; these facts also link them to chordates. Molecular clocks can be used to discern genetic similarities among echinoderms, hemichordates, and chordates.

To understand the importance of the evolutionary transition from invertebrates to vertebrates, think about the extent to which vertebrates have changed the world, occupying every environment on Earth, from the deep oceans to the high mountains. Consider, too, that vertebrates include all fish, amphibians, birds, and mammals. ∎

Chordates: Organisms with the following traits: openings or slits at the region of the pharynx, often called pharyngeal gill slits; a structure called a notochord that runs the length of the organism to support its nerve chord; and a dorsally located nerve chord.

Craniata: A group of vertebrates; as the name implies, craniates have heads supported by skulls that are composed of either cartilage or bone.

Suggested Reading

Ahlberg (editor), *Major Events in Early Vertebrate Evolution.*

Benton, *Vertebrate Palaeontology.*

Questions to Consider

1. Conodonts existed for more than 250 million years, near the start of vertebrate evolution. Speculate on how the evolution of other vertebrate lineages may or may not have contributed their extinction during the Triassic period. What other factors must be considered?

2. Considerable fossil evidence supports that chordates are evolutionarily related to certain non-chordate predecessors. What sort of presently undiscovered fossils—either body fossils or trace fossils—do you think would further illuminate these connections?

Colonization of the Land
Lecture 7

Here's a short summary of what would be needed to make that jump to these terrestrial environments: You need to somehow prevent dehydration; you also need to be able to tolerate temperature extremes. You need to stand up because of greater gravitational challenges; you need to spread your gametes and reproduce. Also, in animals, you need to be able to take in oxygen without dehydrating.

L ife in general adapted to terrestrial environments in the early Paleozoic era, around 500 to 400 million years ago. This was an ecological transition, in which ecosystems moved from marine to land environments.

In the early Cambrian world, the Earth's atmosphere was only about 15% oxygen, as opposed to the current 21%. The land surface was barren, with withered and eroded sediment rather than soil. The transition from sea to land for any form of life in this environment would have presented the potential for dehydration, temperature extremes, and gravitational challenges.

Most paleontologists believe that arthropods were the first animals to make the transition from water to land, largely because arthropods had tough exoskeletons.

Perhaps the most obvious adaptation needed for life on land was the prevention of dehydration. Because all animal lineages early in the Phanerozoic eon evolved originally in full marine environments, freshwater or air would have presented strong selection pressures. In either environment, **osmoregulation** (the process of internal water regulation) would have shifted to block the effects of osmosis in a way to prevent water loss; thus, organisms would have had to develop less permeable membranes.

Temperature is another important factor in the shift from water to land. Water retains heat and maintains a narrower range of temperatures than air, which makes body temperatures easier to control in organisms that are adapted to aquatic conditions. Organisms would have to develop mechanisms for heat retention, such as tough skin.

The third important factor in the move from water to land is greater weight, which means greater gravitational challenges. Water gives organisms buoyancy, which brings certain advantages in energy demands and reproduction. But getting out of the water requires both the strength and the structural integrity to support verticality for plants or movement for animals.

A major glaciation occurred at the end of the Ordovician period that would have lowered sea level and exposed more land for

Tough exoskeletons enabled animals to make the transition from water to land.

organisms living in formerly shallow marine areas. These emergent areas already had colonies of bacteria and archaeans to provide sustenance to newcomers.

The transition of freshwater algae to algae on land probably occurred 475 to 425 million years ago. Terrestrial fungi called ascomycetes were present about 410 million years ago. Lichens, which are symbiotic colonies of algae and fungi, probably started about this time, as well. Land plants, otherwise known as Embryophyta, likely descended from multicellular algae.

Most paleontologists believe that arthropods were the first animals to make the transition from water to land, largely because arthropods had tough exoskeletons. They could both hold in water and provide the structural support needed for moving out of the water. Arthropods developed book lungs, internal respiratory structures, early in the Paleozoic era. Evidence for the earliest animals on land comes from both body and trace fossils. Trace fossils indicate that some animals may have been on land for at least part of their lives before the establishment of land plants.

The earliest terrestrial ecosystems in the Silurian period would have had poorly developed soils with almost no organic matter. There might have been some fungi and lichens. No plants would have been more than about 3 feet tall. Animals, mostly arthropods, would have all been invertebrates, and most of these arthropods would have been predators. In the next period, the Devonian, the addition of insects would change the environment to an even greater extent. ■

Important Term

osmoregulation: The ability of an organism to regulate its internal water "budget."

Suggested Reading

Gensel, P.G., and Edwards, D. (editors), *Plants Invade the Land.*

Shear, *The Early Development of Terrestrial Ecosystems.*

Questions to Consider

1. Of the adaptations plants and animals have in common that allowed them to make the transition from aquatic to terrestrial environments, which do you think is the most important, and why?

2. When some aquatic invertebrates made the evolutionary transition to land environments, was this transition done more easily from fresh-water or salt-water environments? What sorts of fossil evidence would support either hypothesis?

Origins of Insects and of Powered Flight
Lecture 8

In short, insects present the curse of plenty to entomologists who study them, and they clearly represent an astounding sort of success story in terms of their evolution.

At the most basic level, insects are **arthropods**—invertebrate animals with jointed legs. Traits that distinguish them from other arthropods include a body divided into three parts (head, thorax, and abdomen), an exoskeleton, two antennae, and compound eyes.

The oldest fossil insect, *Rhyniognatha hirsti*, dates from the early **Devonian period**, about 400 million years ago. It was originally thought to be a land-dwelling animal, but its later identification as an insect placed insects in terrestrial environments not too long after the colonization of land by other arthropods. *Rhyniognatha* has also been linked to other winged insects, which means that insect flight may have started much earlier than previously supposed.

Insects have been extraordinarily successful from an evolutionary standpoint for a number of reasons. For instance, they have high reproduction rates and huge numbers of offspring; the combination of these two factors ensures high genetic variability in any given population of a species. As a result, insects show rapid responses to any sort of selection pressures.

The group of insects with wings is called Pterygota. The list of orders here includes Ephemeroptera (mayflies), Odonata (dragonflies and damselflies), Dictyoptera (cockroaches and mantises), Dipterans (flies and mosquitoes), Coleopterans (beetles), Hymenopterans (wasps, bees, and ants), and Isopterans (termites). Pterygota can be further divided into the Paleoptera and the Neopteric, of which the Paleoptera are the more primitive. This group includes mayflies, dragonflies, and damselflies, which were common in the Carboniferous period. The most famous dragonfly of this period is *Meganeura*, which lived 300 million years ago and had a wingspan of about 70 centimeters, or close to 2.5 feet.

Several hypotheses have been proposed for the evolution of flight in insects. One of these is the surface-skimming hypothesis. According to this, water-loving insects, perhaps similar to modern mayflies or stoneflies, had either lobes or some sort of modified gills on their thoraxes that assisted them in moving rapidly across water surfaces. The selection for larger structures that became more recognizable as wings would have come from the ability to evade predators or find food or mates more efficiently.

Dragonflies are among the most primitive of flying insects, of the infraclass Paleoptera.

Another theory for the evolution of insect flight is the gliding hypothesis. According to this, non-aquatic insects with similar body parts to the hypothesized skimmers would have been selected because of their improved ability to disperse in terrestrial habitats and, again, to avoid predators, find food, or reproduce.

The initial flight appendages of insects must have had muscle connections in the thorax. Through advances in modern genetics, we now know that the mutations necessary for flight appendages and associated musculature may have been very rapid. The discovery of **Hox genes**, which code for proteins along the axis of an animal, changed our views about how certain appendages on insects might have evolved.

Given that so many insects have aquatic phases in their life cycles, it's not surprising that they have left traces of their activities on stream and lake bottoms. One trace fossil found in Massachusetts, from 310 million years ago, shows where an insect momentarily landed in mud, with no tracks connected to the impression.

The evolutionary transition of insects and insect flight had an enormous impact on the world, especially terrestrial environments. In fact, it's not a stretch to say that the pollination service performed by insects made possible the evolution of primates. ■

Important Terms

arthropods: An invertebrate animal that has a segmented body and jointed legs, such as insects and crustaceans. Trilobites were among the most successful in the Cambrian period.

Devonian period: The period of time from 416 to 359 million years ago; three continents had formed during this time.

Hox genes: Genes that code for proteins along the axis of an animal and are often associated with specific segments.

Suggested Reading

Dudley, *The Biomechanics of Insect Flight*.

Grimaldi and Engel, *Evolution of the Insects*.

Questions to Consider

1. With the scenarios proposed for the evolution of flight in insects in mind, which one do you think is more likely, and what sort of fossil evidence would strengthen this hypothesis?

2. Insects are by far the most evolutionarily successful group of animals. Why did insects diversify more than other terrestrial arthropods, such as arachnids (spiders, scorpions), which originated before insects?

Lecture 8: Origins of Insects and of Powered Flight

Seed Plants and the First Forests
Lecture 9

Nowadays, we take seeds for granted. We barely think about them, whether we're eating peanuts, spitting them out while eating watermelon, or brushing off sesame seeds from our clothes while eating a bagel.... All this we consider as part of everyday normal life with seeds, but it started in the Devonian period, about 370 million years ago.

About 310 million years ago, extensive forests composed of spore-bearing and seed-bearing plants grew in swampy environments along the west side of the then newly formed Appalachian Mountains. Weather conditions along the Appalachians assisted in the formation of broad deltas and swamps, with great amounts of organic material produced by plant communities. These were the vast coal swamps of the Carboniferous period. Most of the plants in these coal-forming forests represented innovations in vascular plants known as **tracheophytes**.

Some spore-bearing plants known as pteridophytes (such as ferns) had evolved during the Silurian period, close to 420 million years ago, and were common during the Devonian period. One well-known pteridophyte is *Archaeopteris*, which probably formed the first true forest. Full-size specimens had trunks more than 3 feet wide and grew to heights of 100 feet. The roots were likewise massive and extended out and into the soil in ways that vascular plants hadn't done earlier.

The selection advantage to greater height in some vascular plants would have been competition for light needed for photosynthesis. Interestingly, as tall plants evolved, they would have created more shade, which means that understory plants would have been selected for shade tolerance.

Another group of tall pteridophytes in the Devonian and the Carboniferous periods were the vascular plants called sphenopsids, which include horsetails and lycopods. The most impressive of the fossil horsetails are *Calamites*, which lived during the Carboniferous period and grew to more than 100 feet tall, towering over modern horsetails. These plants grew clonally, that is, by spreading shoots of the same individual plant laterally, which enabled them to proliferate quickly in terrestrial ecosystems.

Probably the most far-reaching and novel trait that evolved in tracheophytes during this time was the ability to form seeds. Three features had to develop from spore-bearing plants in order to make this transition: (1) a seed coat that serves as protection against desiccation or other damage; (2) nutrients within the seeds, such as starches, oils, or proteins; and (3) an embryo—the potential new plant that will use the nutrients to grow.

Forests changed terrestrial ecosystems on a global scale, altering the landscape, the atmosphere, and the weather.

In plants, the strategy of using spores for reproduction depends on happenstance for fertilization, and because so much is left to chance, spore-bearing plants must release huge numbers of spores into the world. The spores also lack any sort of food supply for growth once they're germinated. In contrast, seed plants depend on internal fertilization.

From the Silurian to the Devonian periods, spores increased in diversity of forms and, eventually, to distinct sizes. The larger megaspore, instead of being let go, stayed in the parent and was fertilized by a microspore. This process, called endospory, is similar to internal fertilization in seed plants.

Seeds helped some land plants adapt to drier conditions, decreasing the need to stay in or close to water sources. Lycopods would have probably grown in dense populations and in more swampy or moist environments. Pteridosperms, the now extinct transitional group, probably grew as upland trees, filling in drier areas.

Forests changed terrestrial ecosystems on a global scale, altering the landscape, the atmosphere, and the weather. The canopies provided by forests also gave vertebrates new opportunities to get out of the water and start moving around on land. ■

Important Term

tracheophytes: A vascular plant that developed during the Carboniferous period.

Suggested Reading

Beerling, *Emerald Planet*.

Kenrick and Davis, *Fossil Plants*.

Questions to Consider

1. Given the statement, "the early evolution of forests may have affected global climate during the Paleozoic era," how could this relationship be examined critically through fossil evidence other than plants?

2. Consider the major evolutionary advantages of land plants that reproduce through closed seeds, rather than through spores. Then why are spore-bearing plants still around today, despite the evolution of seed-bearing plants more than 300 million years ago?

From Fish to 4-Limbed Animals
Lecture 10

Experiments [with Hox genes] are better explaining how what were originally fish fins could later become tetrapod fins in an evolutionary lineage and how such a genetic change could occur relatively quickly. After that, natural selection then took its course, and tetrapod limbs would've been selected favorably under the right circumstances.

In this lecture, we look at the transition from bony fish with lobe-like fins to four-legged vertebrates known as tetrapods. This transition took place in the Devonian-Carboniferous periods, assisted by the evolution of trees and tall plants that provided shade on the forest floor. **Tetrapod** simply means "four legs." It is the group under which all four-legged vertebrates are categorized.

Lobe-finned fish, also known as sarcopterygian fish, are only a small minority of fish today, but a good deal of evidence points to them as the ancestors of tetrapods. The fossil *Tiktaalik* provides a beautiful example of the transition between lobe-finned fish and tetrapods.

Early tetrapods and lobe-finned fish were anatomically much closer than we might think. Lobe-finned fish, so-called because they have four thick, stubby fins, are also members of a larger category of fish called osteichthyans, or bony fish. The tetrapod seems to have started with the sarcopterygian, a group whose survivors today include lungfish and coelacanths. As we might guess from the name lungfish, lungs developed in lobe-finned fish, which allowed them to stay out of the water for extended periods of time.

The four lobed fins of a sarcopterygian fish are its paired pelvic and pectoral fins. These fins are homologous for pelvic and pectoral fins on tetrapods, and many of the same bones are present, as well. One of the greatest evolutionary challenges faced by vertebrates leaving the water was the need to support a greater weight in locomotion on land. To meet this challenge, vertebrates needed to evolve added structural support in their skeletons. The limbs, for example, became lengthened in places, with stronger joints between the bones and stronger muscles. Additionally, the head had to be able to move more freely to facilitate walking and predation; thus, tetrapods had to develop a neck.

Body fossils of lobe-finned fish, such as Eusthenopteron and Panderichthys, show numerous transitional traits between these fish and tetrapods. The pectoral fin of Eusthenopteron, for example, has most of the homologous bones we might expect to find in a tetrapod, such as a humerus, a radius, an ulna, carpals, metacarpals, and phalanges. *Tiktaalik*, mentioned earlier, had a flat head and a body with eyes on the top of the skull, ear notches, shoulders disconnected from the skull, and probably both lungs and gills. Another important tetrapod fossil is *Ichthyostega*, from 370 to 360 million years ago.

© Karen Carr Studio.

A good deal of evidence points to the lobe-finned, or sarcopterygian, fish as the ancestors of the tetrapods, or four-legged land animals.

A spectacular find of probable tetrapod tracks shows many trackways from multiple animals and measurements indicating an animal that was more than 6.5 feet long. Some tracks also have digit impressions showing six to seven digits in the feet. These fossil tracks, from 395 million years ago, have pushed back the possible timing of the transition to tetrapods.

The growth of forests in terrestrial ecosystems would have provided new habitats and many more food resources for the expansion of tetrapods. By the later part of the Devonian period, oxygen content in the atmosphere was probably closer to the 21 percent levels we see today. The next step for tetrapods was to develop enclosed eggs to ensure reproduction during times of drought or in arid environments. With enclosed eggs came other adaptations that caused certain lineages of tetrapods to become the first reptiles during the Carboniferous period, about 320 million years ago. ■

Important Term

Tetrapod: A four-legged vertebrate; the group includes amphibians, reptiles, and mammals.

Suggested Reading

Clack, *Gaining Ground.*

Shubin, *Your Inner Fish.*

Questions to Consider

1. The evolution of forests likely affected the evolution of tetrapods from lobe-finned fish. How could you test this relationship further with either geological or paleontological evidence?

2. Certain anatomical traits had to be favorably selected for lobe-finned fish to evolve for movement on land, such as the strong limbs for support and a neck that allowed for up-and-down movement of the head. What are some other anatomical traits that probably would have helped in this transition, but may not have easily fossilized?

The Egg Came First—Early Reptile Evolution
Lecture 11

One of the main differences [between reptiles and amphibians] is seen in the jaws, as reptiles developed greater biting strength than their amphibian ancestors. This evolution was presumably driven by their main preyed items, which were these probably very large Carboniferous-age insects and other invertebrates.

The expansion of forests across Devonian and Carboniferous landscapes, along with the diversification of insects and tetrapods, changed the face of the land radically. Major changes in the reproduction of tetrapods came next, reflecting changes that must have occurred in their terrestrial environments.

In 1859, the Canadian geologist William Dawson found what is considered to be the earliest known **reptile**, *Hylonomus lyelli*, inside a fossil tree at the Joggins Fossil Cliffs in Nova Scotia. The fossil was found in rocks from the late Carboniferous period, dating from 310 million years ago.

The most basic differences between amphibians and reptiles are not anatomical but ecological. Amphibians have external fertilization and are dependent on water environments throughout their lives. In contrast, reptiles have internal fertilization and can lay their eggs and live in a wide range of environments.

In evolutionary classifications, the group **Amniota** is defined as tetrapods that share a common ancestor with two groups: Synapsida, which includes mammals and mammal-like reptiles, and Sauropsida, which includes all other reptiles, such as dinosaurs and birds. The classification of vertebrates into these evolutionarily related groups, also called clades, changed how we name and discuss these animals. Very different groups of tetrapods previously placed under separate Linnaean categories actually share a common ancestry through their mode of reproduction.

We can broadly classify the clades of amniotes on the basis of three skull types. These are defined by the number of holes, or temporal fenestrae, on the sides of their skulls. A **diapsid**'s skull has two temporal fenestrae on each side, a **synapsid** has one, and an **anapsid** has none. Diapsids and anapsids are both within Sauropsida; modern-day examples include lizards, snakes, alligators, and birds (diapsids) and turtles (anapsids). Modern synapsids are represented by all mammals.

The most basic differences between amphibians and reptiles are not anatomical but ecological.

A number of anatomical features that distinguish amphibians from reptiles are in the skull. One of the main differences is seen in the jaws, because reptiles developed greater biting strength than their amphibian ancestors. Other differences in the skeletons of the earliest reptiles reflect the fact that they spent most of their time on land. The one trait that truly defines amniotes is their amniotic egg, also known as a cleidoic egg.

A fossil called *Casineria kiddi* from the early Carboniferous period shows some transitional characteristics between amphibians and reptiles. It has five digits on both the front and rear feet and limbs that seem better adapted for life on land. Amniote trace fossils from the late Carboniferous generally show impressions of prominent hard claws and five digits, evidence of scales, and indications of a less sprawling gait. Fossil evidence indicates amniotes had already diverged into anapsids, synapsids, and diapsids by about 300 million years ago. Global cooling and a

Reptiles, such as turtles, are able to live independently of water environments and thus have an ecological advantage over amphibians.

© iStockphoto/Thinkstock.

major extinction event about 318 million years ago probably triggered or accelerated this evolution.

The development of enclosed eggs was important in the evolution of vertebrates because it opened more options for tetrapods to live their entire lives and reproduce on land. Terrestrial environments would have offered clear adaptive advantages to any tetrapods that could cut themselves off from water bodies. The development of extensive forests just before the evolution of amniotic eggs suggests that terrestrial ecosystems were likely a driving factor in amniote evolution. ∎

Important Terms

Amniota: An evolutionary classification encompassing tetrapods that share a common ancestor with two groups: Synapsida, consisting of mammals and mammal-like reptiles, and Sauropsida, consisting of all other reptiles.

anapsid: A type of amniote with no holes, or temporal fenestrae, on the sides of its skull. Turtles are part of this group.

diapsid: A type of amniote with two temporal fenestrae on each side of its skull. Modern examples include lizards, snakes, alligators, and birds.

reptile: A tetrapod that fits the behavioral traits for reptiles, such as internal fertilization and the laying of enclosed eggs.

synapsid: A type of amniote with one temporal fenestrae on its skull.

Suggested Reading

Carpenter, *Dinosaur Eggs, Nests, and Babies*.

Sumida and Martin (editors), *Amniote Origins*.

1. Environmental change during the Carboniferous period probably played a role in the natural selection of an enclosed amniotic egg in tetrapods. Why did live birth in amniotes, seen in some modern reptiles, not evolve first, rather than following the evolution of enclosed eggs?

2. The main behavioral difference between amniotes and their more amphibian-like ancestors during the Carboniferous period was that the former laid enclosed eggs on land. Besides anatomical differences, what other evidence might help discern whether any given body fossil from the Carboniferous rocks is an amniote or some closely related non-amniote?

The Origins and Successes of the Dinosaurs
Lecture 12

What's interesting to think about in terms of tetrapod evolution is that bipedal locomotion has also evolved multiple times in very different lineages, including dinosauromorphs, true dinosaurs, birds, and mammals, namely, the hominid line that includes humans.

About 230 to 250 million years ago, in the early to middle **Triassic period**, one lineage of diapsids began evolving the traits now associated with the first dinosaurs. People often think of dinosaurs as being huge, but a large number of dinosaur species ranged from crow-sized to human-sized. More than 500 genera of dinosaurs have been identified thus far, and new genera continue to be found every year. A major revelation in the study of dinosaur evolution is the idea that dinosaurs did not really go extinct; they are still with us as birds.

Perhaps the most important development in the evolution of dinosaurs is represented by the **acetabulum**, an open hole in the hip that allowed the head of the femur to fit into the hip in a ball-and-socket arrangement. This enabled dinosaurs to walk upright or erect, with their legs within a narrow plane in the body. Moving upright probably helped with breathing and enabled speedier locomotion.

Dinosaurs can be split into two clades, mainly on the basis of their hip structures. The **saurischian** ("lizard hip") clade includes all theropod dinosaurs, such as *Allosaurus*, *Velociraptor*, and *Tyrannosaurus*. The **ornithischian** ("bird hip") clade includes all ornithopod dinosaurs, such as Hadrosaura, Thyreophorans, and Ankylosaurs. For each type, the three bones making up the hip are arranged differently.

The split between saurician and ornithischian dinosaurs happened fairly early in their evolutionary lineage, probably about 235 million years ago. Both types of dinosaurs descended from a clade of amniotes called archosaurs. Archosaurs, which are diapsid reptiles, are signified by two important traits: a hole in the skull in front of the eye and teeth held in sockets. Archosaurs included pterosaurs, dinosaurs and their relatives, and some other large animals that went extinct by the end of the Triassic period. Dinosauromorphs were archosaurs that lived during the Triassic period and are regarded as the closest precursors to dinosaurs.

Lagerpeton comes from the middle Triassic period, about 237 to 228 million years ago. It was definitely bipedal; it also had a fourth digit, and its corresponding metatarsal bones are longer than the other toes and metatarsals. This was a relatively small animal, measuring only a little more than 2 feet long. Closer to true dinosaurs is *Marasuchus*, a fast-moving bipedal animal that measured around 4 feet.

Most of the earliest known dinosaurs are saurischians and theropods, although a few are primitive ornithischians. *Eoraptor* is largely accepted as the oldest known dinosaur. Both its femur and its ankle were adapted for a more erect posture, and its feet had three prominent digits. The earliest ornithischian dinosaurs include two significant specimens, *Pisanosaurus* and *Eocursor parvus*.

© Karen Carr Studio.

Environmental conditions favored the evolution of dinosaurs to huge sizes and to domination of the Mesozoic landscape.

Environmental conditions favored the evolution of dinosaurs to huge sizes and to domination of the Mesozoic landscape. At the end of the Triassic period, a mass extinction took place, probably initiated by the breakup of the supercontinent Pangaea, which brought on global warming and the extinction of many seed plants. Numerous archosaurs died out, opening up ecological niches for dinosaurs. Throughout the remainder of the Mesozoic era, dinosaurs occupied every major terrestrial habitat—underground, on the ground, and in the trees. ∎

Important Terms

acetabulum: An open hole in the hip of dinosaurs that allowed the head of the femur to fit into the hip in a ball-and-socket arrangement.

ornithischian: Literally, "bird hip"; one of two clades into which dinosaurs are divided; includes all ornithopod dinosaurs, such as Hadrosaura, Thyreophorans, and *Stegosaurus*.

saurischian: Literally, "lizard hip"; one of two clades into which dinosaurs are divided. Includes all therapod dinosaurs, such as *Allosaurus*, *Velociraptor*, and *Tyrannosaurus*.

Triassic period: The period of time from 251 to 200 million years ago; the time when the supercontinent Pangea started to break up into the continents we know today.

Suggested Reading

Fraser and Henderson, *Dawn of the Dinosaurs*.

Martin, *Introduction to the Study of Dinosaurs*.

1. The earliest dinosaurs in the late Triassic period were relatively small compared to their descendants in the remainder of the Mesozoic. What evolutionary factors could have later resulted in the natural selection of extreme sizes in some dinosaurs, while other dinosaurs also evolved to be as small as some modern songbirds?

2. Dinosaurs share a common ancestor with pterosaurs and crocodiles, but both of these groups are not dinosaurs. How did their evolutionary paths result in behavioral and anatomical differences between these three groups of animals?

Marine and Flying Reptiles
Lecture 13

In these extraordinary fossils from the early Jurassic of Germany, from about 190 million years ago, we see baby ichthyosaurs exiting the birth canal of the mother ichthyosaur with their tails coming out first. This behavior is a clear adaptation for living and reproducing in the water, as modern whales and dolphins show exactly the same sort of position of their young during live birth.

The pterosaurs were evolutionary cousins and contemporaries of the dinosaurs in the Mesozoic era. Some of the most important Mesozoic vertebrate fossil finds of the 19th century were made by Mary Anning, who helped discover the first known specimen of the marine reptile *Ichthyosaurus* and the first known plesiosaur.

If you recall, the major skull types for amniotes that we discussed in previous lectures are synapsid, diapsid, and anapsid. A fourth skull type that developed in some marine reptiles, such as ichthyosaurs and plesiosaurs, is a euryapsid skull.

Probably the most intriguing marine reptiles from an evolutionary standpoint were the ichthyosaurs. They were active eaters of seafood, they had limbs that were modified into flippers, and they had streamlined bodies. One species of ichthyosaur developed the largest known eyes of any vertebrate, about a foot in diameter, presumably an adaptation to reduced light conditions at great ocean depths. Ichthyosaurs also had live births, representing the first instance of this evolution in amniotes.

Plesiosaurs are another famous group of marine reptiles from the Mesozoic era. The fins of plesiosaurs, which look very much like modified hands and feet, show their relationship to land-dwelling ancestors. Most of these animals probably ate fish and mollusks, and their well-developed fins meant that they were active swimmers. The short-necked plesiosaurs included some of the largest and most important predators of the Mesozoic seas. One of these was the late Jurassic *Liopleurodon*, which reached lengths of about

30 feet. Competing for seafood in Cretaceous oceans were the awesome predators called mosasaurs, some of which were more than 55 feet long. Another group of reptiles distantly related to mosasaurs yet also adapted to marine environments during the Cretaceous period includes snakes. The largest sea turtle that ever lived, *Archelon ischyros*, is from about 70 million years ago and was 15 feet long.

A mass extinction at the end of the Cretaceous period, probably caused by a meteorite impact, wiped out dinosaurs, mosasaurs, pterosaurs, and many other animals.

Before birds, the world's skies were filled with a wide variety of pterosaurs—more than 100 fossil species at last count. Pterosaurs were apparently the first vertebrates with powered flight. The oldest pterosaur, *Eudimorphodon*, dates from 210 to 220 million years ago. This gull-sized pterosaur had an enormously elongated fourth digit and a novel bone in its wrist that helped to support wing membranes for flight.

An interesting example of convergent evolution in pterosaurs is provided by hair-like projections on their skin, which probably served as insulation. This hair, along with metabolic needs for sustained flight, suggests that pterosaurs may have evolved warm-bloodedness, or endothermy, independently of mammals. Sizes of pterosaurs ranged from the tiny *Nemicolopterus crypticus*, with about a 10-inch wingspan, to the giant *Quetzalcoatlus*, with a wingspan of about 36 feet. A good number of pterosaur trackways have been found, showing that these animals walked on all fours.

A mass extinction at the end of the Cretaceous period, probably caused by a meteorite impact, wiped out dinosaurs, mosasaurs, pterosaurs, and many other animals. Dust kicked up by the impact and smoke from forest fires would have shut down photosynthesis for several years. Global temperatures also would have plummeted. The only organisms that would have survived such a shock were those of small size with burrowing abilities. ■

Suggested Reading

Everhart, *Oceans of Kansas*.

Unwin, *The Pterosaurs*.

Questions to Consider

1. Amniotes diversified considerably during the Mesozoic era, resulting in the evolution of ichthyosaurs, plesiosaurs, mosasaurs, sea turtles, and pterosaurs. How did sea-level changes during the Mesozoic era probably affect the diversification and extinctions of these marine reptiles, and what other factors must be considered other than sea level?

2. Pterosaurs were the first flying vertebrates, and probably adapted to a wide range of feeding habits. Given any pterosaur body fossils or trace fossils, how could you infer their feeding behaviors?

Birds—The Dinosaurs among Us
Lecture 14

All in all, one could not ask for a much better example of a blend between dinosaur and bird than *Archaeopteryx*. What continually astonishes paleontologists about this fossil species is how we're still learning about it more than 150 years after its discovery and following intense study and considerable scientific debate.

The evolution of birds differs from that of pterosaurs on the basis of two major points: (1) Birds almost certainly evolved from a lineage of flightless theropod dinosaurs, and (2) pterosaurs went extinct at the end of the Cretaceous period, whereas birds survived and thrived.

Paleontologists are now quite certain that birds descended from small theropod dinosaurs during the middle to late Jurassic period, probably about 165 to 155 million years ago. *Archaeopteryx*, which we saw in Lecture 1, is about 150 to 145 million years old. It had feathers and a number of other traits showing that the divergence of theropods into avian and non-avian categories had already taken place.

At a minimum, birds have (or had and later lost) the following traits: feathers, a reduction in the number and sizes of the

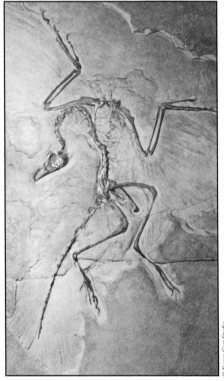

© Karen Carr Studio.

An *Archaeopteryx* fossil.

46

vertebrae in their tails, forearms that are far more than 90% the length of their humerus, forelimbs that are more than 120% longer than their hind legs, and three forward-pointing digits on their feet and one digit pointing backwards. The first four of these traits are adaptations for flight. The backward-pointing digit is an adaptation for perching on tree branches.

Archaeopteryx had a tail with 21 to 22 vertebrae, which made its tail long for a bird but short for a theropod dinosaur. It had a furcula, or wishbone, which is found in all birds but only a few types of theropods. Its breastbone, or sternum, was poorly developed, unlike those of flighted birds. The skull of *Archaeopteryx* had teeth in both jaws, which modern birds do not. CAT scans done on skull interiors of *Archaeopteryx* have also revealed that it was more bird-brained than reptilian-brained. A recent study of growth lines showed that it had a lower calculated growth rate than that of a typical bird.

How would the world be different if birds had not evolved from dinosaurs?

In the late 1990s, paleontologists began to find numerous small, feathered theropod dinosaurs from the early Cretaceous period. One of the most spectacular of these was the theropod *Microraptor*, which had feathers on all four limbs and its tail. The oldest feathered dinosaur is *Anchiornis huxleyi*, dating from 160 to 155 million years ago. Even *Velociraptor*, a theropod from about 70 million years ago, has been shown to have had attachments for quills on its arms, similar to what can be seen on the forearms of vultures. Feathers, which are basically heavily modified scales, first evolved in theropod dinosaurs before fully flighted birds evolved. Feathers probably developed for insulation and to identify species and potential mates.

Three hypotheses have been proposed for the evolution of flight. The first of these is the ground-up hypothesis, which states that small, fast-running theropods evolved flight from the selection of flight-adapted traits on the ground, such as escaping predators and catching prey. The trees-down hypothesis states that small theropods were tree climbers that evolved flight first by gliding, then by flapping their arms. The assisted-running hypothesis synthesizes aspects of the preceding two and is supported by experiments done on living birds.

How would the world be different if birds had not evolved from dinosaurs? Obviously, without songbirds, it would be a lot quieter. We would also experience a boom in the populations of flying insects. Perhaps most important, we would not have instruments for seed dispersal of flowering plants. ■

Suggested Reading

Chiappe, *Glorified Dinosaurs*.

Shipman, *Taking Wing*.

Questions to Consider

1. When paleontologists make the statement "birds are dinosaurs," how does it compare to the statement "humans are primates"?

2. Review the three hypotheses for the origin of flight in feathered dinosaurs. Based on both the fossil record and experimental work with modern birds, which of the three do you think is most probable, and why?

The First Flowers and Pollinator Coevolution
Lecture 15

Although some of the results may seem subtle, the overall impact of flowering plants and their pollinators was perhaps no less important for the overall future of life on land than, say, the first forests had been, back when they transformed the atmosphere and the landscapes of the planet.

Flowering plants, known as **angiosperms**, number about 250,000 species and account for about 80% of all land plants. The animals that pollinate these plants—insects, birds, and even mammals—are arguably even more diverse. The origins and diversification of birds overlapped with the origins of flowering plants, which occurred at least 130 million years ago and was quickly followed by the origins and explosive diversification of many pollinating insects.

Flowering plants reproduce by using **pollen**, which contains male sperm. Somehow, this pollen must be transported to the female parts of the same species of plant, which contains ovules. If the sperm fertilizes the ovules, seeds are produced and fruits grow to surround these seeds.

Flowering plants offer terrestrial animals pollen, nectar, and fruit to entice them to lend a hand in plant reproduction. Insects are attracted to plants as sites for reproduction, as shelter from predators, and as food. In fact, many insects have evolved forms of camouflage to better fit in with the visual appearance of plants and take advantage of these benefits.

The production of nectars is interesting because it represents a blatant adaptation of plants for attracting insects. Nectars, which consist mostly of simple sugars, were easy for the first flowering plants to make. For the pollinators, these simple sugars supply energy, maximizing their ability to fly from flower to flower and assist with pollination.

Some Jurassic gymnosperms called Gnetales are regarded as precursors of angiosperms, in part because modern Gnetales have special tissues called vessel elements that aid in transporting water in vascular plants. Such vessel elements are absent from all other gymnosperms, such as cycads and conifers, but they are in angiosperms. However, this shared trait does not necessarily represent evolutionary relatedness.

The oldest undoubted angiosperm fossils are from three species of the same genus, *Archaefructus*, dating from about 125 million years ago. Interestingly, *Archaefructus* was a water-dwelling plant and may have been related to lilies, which has led to speculation that the early evolution of flowering plants may have been in watery environments. One Jurassic plant, *Schmeissneria*, has been proposed as a possible angiosperm ancestor or even as an angiosperm itself. Chemical fossils called oleananes link with angiosperms and suggest an even deeper origin for flowering plants, going back more than 250 million years.

It might seem surprising that flowering plants survived the end-Cretaceous mass extinction. However, by that time, flowering plants had undergone at least 60 million years of adaptive radiation. Further, flowering plants benefited greatly from their mutual alliances with virtually all types of small terrestrial animals and insects that also survived the extinction.

The important pollinators of today that probably

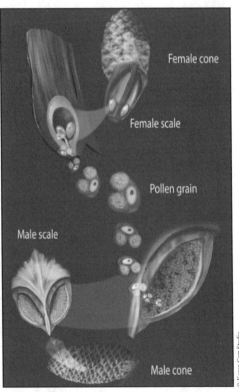

© Karen Carr Studio.

Flowering plants, which reproduce using pollen, emerged about 130 million years ago.

coevolved with the earliest flowering plants include wasps and bees, butterflies and moths, beetles, ants, hummingbirds, bats, honey possums, and lemurs. Many of these insect pollinators were present during the Mesozoic era and become noticeable in the fossil record after flowering plants exploded in diversity. The relationship between angiosperms and potential pollinators is so close that it's difficult to·tell whether this coevolution was driven first by plants or insects. ■

Important Terms

angiosperms: Flowering plant; characterized by enclosed seeds.

pollen: A powder that contains the male gametes of seed plants.

Suggested Reading

Grimaldi and Engel, *Evolution of the Insects*.

Taylor, T.N., Taylor, E.L., and Krings, *Paleobotany*.

Questions to Consider

1. Non-flowering plants had evolved and thrived for more than 200 million years before the origin of flowering plants (angiosperms). Given what you know about seed plants and their animal cohorts, what do you think may have triggered the initial evolution of angiosperms from their gymnosperm ancestors during the Mesozoic era?

2. Many animals, such as certain flying insects, birds, and mammals, likely coevolved with angiosperms during the Cretaceous Period. What future fossil finds do you expect may provide more information about this coevolution?

Egg to Placenta—Early Mammal Evolution
Lecture 16

Better respiration, better digestion, better ability to withstand cold temperatures, better sense organs for operating at night, better jaws, and better teeth that helped some of them be able to eat a wider variety of food items—all those adaptations and exaptations prepared mammals to live in nearly every terrestrial habitat, with a bewildering variety of forms and behaviors emerging all throughout the last 65 million years.

Mammal-like reptiles arose from **synapsid** reptiles in the Permian period, about 290 to 250 million years ago. In other words, mammals and their ancestors appeared and began to evolve at almost the same time and for some of the same reasons as the dinosaurs. In fact, at least one Cretaceous mammal, *Repenomamus robustus*, preyed on dinosaurs.

The general characteristics of mammals include hair, mammary glands, relatively large brains compared to their body sizes, and long care for their offspring. Mammals also have several bones that reflect an evolutionary connection to their more reptilian ancestors. The three modern types of mammals are monotremes, which are egg-laying mammals; marsupials, which have live birth, but their embryonic offspring are nurtured in a pouch for an extended time; and eutherians, which are placental mammals.

Mammals are, most basically, synapsids, a group identified by a single temporal fenestra on each side of the skull. A general trend in early synapsid skulls was an enlargement of the temporal fenestrae. This was probably the result of selection for larger jaw muscles that attached the bones around the holes in the skull. Such jaw muscles enabled chewing food, as opposed to swallowing it whole.

Synapsids in general were the ruling reptiles of the Permian until a mass extinction at the end of the period. Among the synapsids that survived this mass extinction was a group called **therapsids** that spread worldwide. These animals show a number of traits associated with mammals, such as differentiated teeth, stronger jaws, modifications of the jaws to enable better hearing, and an upright posture.

One group of therapsids, cynodonts, shows numerous modifications to their jaws that indicate more efficient chewing. For example, the lower jawbone became one solid piece rather than several articulated bones, as we find in pre-therapsid reptiles. Two of the reptile jawbones moved up to the inner ear. Another important shift was the full development of a secondary palate in the roof of the mouth, enabling these animals to eat more food by separating breathing tubes and feeding cavities.

Elephants are the only surviving mammal family of the order Proboscidea.

As synapsids went from the more sprawling posture associated with a normal reptile to a more upright one, their limbs also lengthened, tails shortened, and hipbones got smaller. These shifts are thought to reflect the energy demands of increased brain size, metabolism, and activity, all of which are likely associated with **endothermy**. Endothermy itself may have developed as one adaptation for nocturnal activities.

The early Jurassic *Morganucodon* exhibits the mammalian trait of diphyodonty (i.e., teeth are not replaced endlessly if they are lost). This aided in improved occlusion and suggests suckling behavior. The most successful of Mesozoic mammal groups were multituberculates, which evolved by the late Jurassic period and lived through the Eocene epoch of the Cenozoic era.

Various synapsids and therapsids went extinct at the end of the Permian period, leaving only the cynodonts. Thus, ecological niches were likely opened for cynodonts in the early Triassic, which led to adaptations that took their lineages down the evolutionary path to mammals. Another mass extinction at the end of the Cretaceous period opened niches for both mammals and birds in terrestrial ecosystems. This, in turn, promoted diversification of mammals in the Cenozoic era. ■

Important Terms

endothermy: Physiology in which an organism generates its own internal body heat; sometimes called "warm-blooded.

synapsid: One of three classifications within the group Amniota; a synapsid's skull has one temporal fenestra. Modern synapsids are represented by all mammals.

therapsids: A group of reptiles of the Permian period that show a large number of traits associated with mammals, such as differentiated teeth, stronger jaws, modifications of the jaws to enable better hearing, and an upright posture.

Suggested Reading

Kemp, *The Origin and Evolution of Mammals*.

Kielan-Jaworowska, Cifelli, and Luo, *Mammals from the Age of Dinosaurs*.

Questions to Consider

1. Some important anatomical features distinguished mammals from their synapsid reptile ancestors. How did these traits, such as modification of some jawbones into tiny bones of the inner ear, reflect novel adaptations that were shaped by natural selection?

2. Think about how mass extinctions relate to mammalian evolution over the past 250 million years. For instance, how did the end-Permian period extinction 250 million years ago differ in its effect on mammalian evolution from the end-Cretaceous extinction 65 million years ago?

From Land to Sea—The Evolution of Whales
Lecture 17

This size range that you have in whales is analogous to how dinosaurs ranged from the crow-sized *Microraptor* to long-necked sauropod dinosaurs that exceeded 50 tons. Nevertheless, the range of size differences in cetaceans is really remarkable, and it's unique among any group of animals.

After the mass extinction at the end of the Cretaceous period, mammals began to fill ecological niches left by dinosaurs, marine reptiles, pterosaurs, and many other animals. The biggest and perhaps most surprising example of the radiation of mammals was their transition into the oceans in the evolution of whales.

For some time, whales were thought to have descended directly from mesonychids, which were carnivorous, land-dwelling, hoofed mammals that lived during the early Eocene epoch, about 35 to 40 million years ago. However, later fossil discoveries, including that of a *Pakicetus*, placed whales in the category of artiodactyls (even-toed, hoofed mammals). *Pakicetus* had middle ear bones that enabled it to hear underwater and an anklebone that compares well with those of modern artiodactyls.

Modern cetaceans retain many vestigial traits and behaviors that reveal their evolutionary origins from land-dwelling placental mammals. Hind limbs and hipbones are greatly reduced, as in snakes, but are still present. Whales also have homologous bones in their front flippers that match those in the forelimbs of other mammals. Perhaps the most overt behavioral trait demonstrating the ancestry of whales from land-dwelling mammals is the up-and-down motion they have while swimming. This movement reflects the same spinal flexion as running in mammals.

Modern cetaceans are divided into two main groups: odontocetes, or tooth whales, and mysticetes, or baleen whales. Tooth whales are mainly predatory, while baleen whales strain plankton from the water. Baleen whales, including the blue whale, evolved to become the largest animal of all time, measuring more than 100 feet long and weighing more than 165 tons.

The fossil record for the earliest whales and their close relatives is quite good. For example, *Indohyus*, dating from about 48 million years ago, was a small hoofed mammal, similar to a deer in form. *Indohyus* had relatively dense bones, which would have helped it stay submerged in water for extended periods of time to avoid predators. *Pakicetus*, mentioned earlier, was a carnivorous hoofed mammal with differentiated teeth, but it was likely amphibious. A skeleton of *Maiacetus*, from about 47 million years ago, has been found with a fossil fetus inside it. The position of the fetus, head downward, indicates that primitive whales may have returned to land to give birth.

Recent experiments on dolphins demonstrate that they can recognize themselves in mirrors as individuals.

Genetic tests have shown that whales and artiodactyls were clearly related to each other. Similarities in DNA in extant whales show a closer affinity to hippopotamuses than to other mammals. Molecular clocks indicate probable divergence of whales from their artiodactyl ancestors at about 60 million years ago.

The selection of larger brain sizes in dolphins is, of course, correlated with more complex intelligence and behavior. Recent experiments on dolphins demonstrate that they can recognize themselves in mirrors as individuals. Dolphins also use complicated auditory communications, an adaptation that is connected to social interactions and cooperation. Studies of fossil dolphins and their brain cases have revealed two probable phases in the development of dolphin intelligence, at about 39 and 15 million years ago. These evolutionary transitions in cetaceans are pertinent to better understanding human evolution and how convergent evolution could have occurred through natural selection for larger brains in very different animals, one in the sea and the other on land. ■

Suggested Reading

Berta and Kovacs, *Marine Mammals*.

Prothero and Foss (editors), *The Evolution of Artiodactyls*.

Questions to Consider

1. Consider the currently accepted hypothesis for the origin of modern whales. What do you regard as the best available evidence from the fossil record supporting this hypothesis?

2. Think about the evolutionary factors that resulted in the selection of relatively large brain size (and intelligence) in some lineages of whales, such as dolphins. How does this represent a type of convergent evolution with primate lineages leading to humans, despite their evolution in separate environments (marine and terrestrial)?

Moving on Up—The First Primates
Lecture 18

> The anthropoid features that we see today—larger brains, an accentuation of the visual sense, a closure of the eye orbit behind it, and color vision—those features seem to have originated as a consequence of the origin of the anthropoids, not as causes of them. I say that because they all originated after the anthropoids originated.

In this lecture, we discuss the transition from early primates to **anthropoids**, which include hominoids (living apes and their fossil ancestors, as well as humans) and New World monkeys. New World monkeys live in the Americas and include tamarinds, marmosets, spider monkeys, squirrel monkeys, and others. Old World monkeys live in Europe, Africa, and Asia and include baboons, macaques, langurs, and colobus monkeys. All of this diversity originated from one ancestor sometime during the Eocene epoch.

Georges Cuvier was a prominent French scientist and opponent of evolutionary ideas in the early 19th century. However, Cuvier made important contributions in the fields of anatomy and paleontology, including his work in the Montmartre gypsum beds. These beds, located in Paris, were composed of sedimentary layers dating from the Eocene epoch. Here, Cuvier found a fossil primate that he named *Adapis*, believing it to be some sort of cow-like herbivore. Today, we know this fossil was a primate by virtue of its teeth and skull. In general, we can place mammals within different classes based on their teeth, although with early mammals, this process can be difficult.

Genetics has also helped to clarify relationships among early mammals. For example, we now know that primates are closely related to some kinds of small insectivores, such as bats, rodents, dermopterans (flying lemurs or colugos), and tree shrews.

The diversification of early primates and closely related mammals occurred rapidly. One example of early primate-like mammals is the group plesiadapids. The skeletons of these animals seem to converge on primates, but their crania are quite different from those of primates.

The fossil record of Cuvier's primate, *Adapis*, is so dense in North America that we can trace its evolution over about 8 million years of the Eocene epoch. We can see, for example, the gradual increase in tooth size, as well as the divergence in tooth size of three different lineages.

Probably the most famous of the adapids to be discovered so far is the fossil called *Darwinius*, which tells us a number of details about adapid primates. One particular skeleton, nicknamed Ida,

Lemurs, such as the ring-tailed lemur, are similar to the extinct primate family of Adapids.

was identified as a juvenile from her tooth development. Her body growth rate seems to be very similar to that of living primates.

Another candidate for the ancestor of anthropoids is a primate from the family Omomyidae. These animals are similar in their skulls to living tarsiers. In contrast, Adapids are similar in their skulls to living lemurs. Both tarsiers and lemurs (prosimians) tend to be nocturnal, are insect predators, and are generally solitary. Monkeys and apes, however, are different; they're mostly diurnal, eat a mixed diet, and tend to live in complex social groups.

Primate fossils have been found in the Faiyum Depression in Egypt since the early 20th century. The species *Afradapis*, for example, shows many convergent features shared by adapoids and anthropoids, particularly in the jawbone. An important anthropoid found in Southeast Asia is *Eosimias*, dating from about 43 million years ago.

The fossil record of the early anthropoids leads to the hypothesis that they originated from a tarsier-like ancestor somewhere in Asia and dispersed into Africa relatively quickly. New World monkeys may have traveled to South America by floating on trees or other vegetation down the rivers of Africa. Such "rafts" may have had enough resources to keep monkeys alive as they crossed what was once a much smaller Atlantic Ocean. ∎

Important Term

anthropoids: A group of primates that includes monkeys, apes, and humans.

Suggested Reading

Beard, *The Hunt for the Dawn Monkey*.

Questions to Consider

1. The known early primates were all the size of small monkeys or smaller. We still have tiny primates, but they have been joined by large terrestrial monkeys, apes and lemurs. Why might recent primates include this range of large body sizes that early primates did not have?

2. Primates depend on learning many things from the members of their social groups. How might this reliance on learning have affected the growth and development of primates?

Apes—Swinging Down from the Trees
Lecture 19

The kinds of characteristics that we used to think were really distinctive similarities of humans and apes may have, in many cases, evolved in parallel. But there are other characteristics of the apes, including their long life history, their behavioral complexity, and their dental adaptations, that are true links to humans and to more ancient apes.

Old World monkeys are the closest relatives of the apes, but living apes share a number of features that Old World monkeys don't have. Many of these features are related to locomotion; others are related to dental characteristics and life history characteristics, such as a long rate of maturation and a relatively long lifespan. The group to which living apes, their fossil relatives, and living humans belong is the hominoids.

A cladogram places the origins of all living great apes and humans at about 10 to 13 million years ago. Gibbons, which are called lesser apes, diverged substantially earlier than the great apes, about 16 to 20 million years ago. That divergence represents the common ancestor of all living hominoids. Cercopithecoids, the Old World monkeys, diverged from our lineage 22 to 26 million years ago. Humans and chimps shared an ancestor about 4 to 7 million years ago; gorillas and humans, about 6 to 9 million years ago; and orangutans and humans, about 10 to 13 million years ago.

The Miocene period, in which apes diversified, was a different world than the one we live in today. The weather was substantially warmer and milder, and forests covered most of the tropical and temperate world. The biomes of east and central Africa were not separated to the extent that they are today; thus, ape habitat was dispersed across most of the Old World. In addition, the seas that once separated Asia from Africa and Europe from Asia were shrinking, allowing primates to disperse and diversify.

The earliest ape for which we have good fossil evidence is *Proconsul*, which lived between 22 and 16 million years ago. *Proconsul* was a quadruped with a pronograde posture, meaning that its back was horizontal when it was walking. It had the thick-enameled teeth of apes, developed for grinding rather than scissoring. Another early primate from east Africa was *Morotopithecus*. It had a lumbar spine that was stiff and vertical, suggesting that it was more of a climber than a runner. *Kenyapithecus*, from about 14 million years ago, has a dental pattern that is similar to that seen in living apes, making it a good candidate for being the first ape to disperse out of Africa.

Among the outstanding questions in ape evolution is why some apes evolved toward walking and others toward climbing.

Dryopithecus, from 13 million years ago, lived in Europe. European apes probably ate as many kinds of foods as today's apes do. During the middle to late Miocene, the Old World monkeys evolved two distinct branches based on diet. The colobines specialized in eating leaves and developed a complex stomach to enable them to digest leaves in coordination with microbes. The cercopithecines ate fruits and hard seeds. The advantage that these monkeys had over the apes was their short birth intervals; thus, over the course of the late Miocene, the smaller apes begin to decline.

Among the outstanding problems in ape evolution is the issue of locomotor diversity. Why did some apes become specialized toward quadrupedal movement, while others became specialized toward climbing and hanging from branches? The fossil *Ardipithecus*, from 5.5 to 4.3 million years ago, gives us some evidence about locomotion. From the bones of its hands, we know that it was not a knuckle-walker, and it carried its spine more vertically than most of the great apes do today. This ape may represent a close relative of the ancestor of hominids and, later, humans. ■

Suggested Reading

Taylor, *Not a Chimp*.

Walker and Shipman, *The Ape in the Tree*.

Questions to Consider

1. Why might it be that the giant *Gigantopithecus* in south China, which survived at least until 300,000 years ago, persisted so long when almost all the other Miocene apes became extinct before 5 million years ago?

2. Apes have long life histories, with very slow potential for population growth. Yet they were very successful at colonizing Europe and Asia during the Miocene. How much of their success do you think was attributable to their environments, and how much to their behavioral and anatomical adaptations?

From 4 Legs to 2—The Hominin Radiation
Lecture 20

That commitment to an obligate form of bipedality means that you're compromised in many respects. The first compromise to strike us was the increased risk of predation. Just the fact that now you're out on your own—you might look a little bigger to a predator because you're taller—but if you stray a little bit far from the trees … you're at extreme risk of one of those predators getting you.

Recall that the hominids are the branch of primates that are closely related to humans and more distantly related to any of the living apes. Our branch originated between 4 and 7 million years ago. Hominids lived only in Africa up until 2 million years ago.

Ardipithecus, which we saw in the last lecture, had some characteristics that suggest it is either a possible ancestor of the hominids or a close relative to our common ancestor with chimps and gorillas. Key among those features are the teeth, particularly its canine teeth. Canine teeth allow carnivores to eat meat, but when they are displayed, they can also serve as a threatening signal. This tells us that late Miocene apes, such as *Ardipithecus*, may have had a social organization and may have been communicating a bit like we do and a bit differently from living great apes.

The most well known of the bipedal primates is the gibbon. Gibbons have extraordinarily long arms, and they're especially good at swinging in trees, but when they're on the ground, these long arms are almost useless. When a gibbon is

Early hominids used their canine teeth for eating meat and for threat displays.

on the ground, it runs bipedally, using its arms to balance itself. Humans today exhibit what is called **obligate bipedality**, which involves many elements in our skeletons. The fossil that most clearly demonstrates the differences here is Lucy, who lived about 3.2 million years ago. Lucy was probably around 15 to 20 years old, stood 3.3 feet tall, and weighed about 80 pounds. She is small compared to living humans, but her bones align with ours in the characteristics that show bipedal locomotion.

> Several features of hominin bones show adaptations to bipedality, including a curvature of the hipbone. This shape keeps the body upright when we're standing on one foot, as we do with every step we take.

Several features of **hominin** bones show adaptations to bipedality, including a curvature of the hipbone. This shape keeps the body upright when we're standing on one foot, as we do with every step we take. The curved hipbone also ensures that we move forward as we walk. Other anatomical adaptations for bipedality include an enlargement of the glutcus maximus muscle, a tibia that comes straight up from the foot, and an angulation at the knee that enables the hip joint to remain to the side of the body when standing on one leg.

Trace hominin fossils are relatively rate, but among the most famous are the footprint trails from a site in Tanzania, dating to about 3.6 million years ago. These clearly show two individuals walking side by side, and their spacing suggests that the larger hominin may have been walking more slowly so that the smaller one could keep up.

The first australopithecine fossil found was from a cave called Taung in South Africa. It is part of the face and jaw of a young child, together with the natural cast of the inside of the child's skull. The development of the teeth here is much more human-like than that of chimpanzees. The talon marks of the eagle that originally ate this juvenile primate are evident on its skull.

Over the course of several million years, hominins evolved into many different types. There were probably more than 7 or 8 species of hominins that fell into different niches based on their diet. The relatively large molars, along with isotopic studies of bones, tell us that early hominins ate nuts, some parts of grass, and insects. ■

Important Terms

hominin: Clade of primates that includes all species more closely related to humans than to other primates. Includes *Homo* and *Australopithecus*.

obligate bipedality: The anatomical requirement to walk only on the lower limbs; a defining trait for classification as a human.

Suggested Reading

Gibbons. *The First Human: The Race to Discover Our Earliest Ancestors.*

Johanson and Wong. *Lucy's Legacy.*

Questions to Consider

1. Primates that spend a lot of time on the ground tend to have larger groups than those that stay in the trees. How can this affect our understanding of the origin of bipedality in our lineage?

2. Humans have colonized almost all habitats on Earth, but many required the advanced cultural skills of Homo. What part did bipedality play in our success as a colonizing species, and why do you think Australopithecus had trouble leaving Africa?

First Humans—Toolmakers and Hunter-Gatherers
Lecture 21

It's quite possible that the success of our lineage might have had to do as much with our immune system as it did with the kinds of stone tools we could make or the brain size that we've got—or all of these things might have worked together, so that in order to evolve a bigger brain size, you first had to conquer the diseases that were holding you down.

Between 1.5 and 2 million years ago, at least three species of humans lived in various parts of Africa. One of these species is *Homo erectus*, which was probably our ancestor. The others are *Homo habilis* and one of the robust australopithecines.

The Olduvai Gorge is one of the most well-known hominin fossil sites, dating from about 1.9 million years ago. Oldowan technology (technology that hearkens from this area) consists of pebbles that have been struck hard with another rock to produce sharp-edged flakes that could be used for cutting. Rocks could also be used to smash bones in order to get to the marrow inside. Archaeological research has shown that hominins may have preferred certain types of raw material for their tools and carried such stones with them as they traveled.

One possible candidate for the toolmaker in the Olduvai Gorge, known as Olduvai Hominid 7, became the basis of the species *Homo habilis*. It is substantially advanced over australopithecines in at least two ways: Its thick fingertips indicate that it could apply more force to grip an object strongly, and its brain was about 25% larger than that of any australopithecine.

A skull of *Australopithecus* from 2.8 million years ago shows a relatively small brain size, a sloping face, and a nose that doesn't project outward from the face. In contrast, a *Homo habilis* skull from 1.65 million years ago shows a larger brain, a more vertical face, and smaller teeth. It also has more development in a region of the brain called Broca's area that corresponds to language production in modern humans.

The earliest stone tools are 2.6 million years old and are known to have been used to process meat. What we don't know, however, is who these early toolmakers were. Nevertheless, the shift to a higher-quality diet that probably included meat had an influence on the evolution of our jaws and can be traced in our genetics. A gene that expresses a protein in the muscle of the jaw is active in all other primates but not humans, suggesting that our jaw muscles were reduced to a substantial degree. Such **"pseudogenes"** also appear in red blood cells and may be related to immunity to different types of pathogens.

© Goodshoot/Thinkstock.

Greater height is a human adaptation that allows travel by foot over longer distances.

One of the most important aspects of our lineage compared to Australopithecus is the size of our bodies. The first evidence of a large body adaptation in *Homo* is seen in skeleton WT 15000, the Nariokotome boy. It's about 1.6 million years old, but at age 10 or 11, this boy stood about 5 feet, 3 inches tall. As an adult, he might have been closer to 5 feet, 10 inches. Such height would have enabled this human to travel farther away from water sources to forage without getting dehydrated.

Another consequence of larger bodies is that the pelvis remained short, but babies became larger, making the birth process more difficult in humans than in other primates. One of the ways that humans have adapted to the birth process is by delaying the growth of our brains until after we're born.

A cranium from Dmanisi, in Georgia, reveals that cooperation may have been evolving in early humans. This skull is from an older individual who lost all of his or her teeth, which means that this person would have required help in getting food. That signal of culture will be key to our next transition to modern humans. ■

Important Term

pseudogenes: Nonfunctional remains of a formerly functional gene.

Suggested Reading

Zimmer, *Smithsonian Intimate Guide to Human Origins*.

Questions to Consider

1. Humans have a tremendous reliance on our ability to make and use tools. What other kinds of animals depend so strongly on manipulating objects aside from their own bodies?

2. Australopithecines were successful for up to 4 million years, but disappeared from the fossil record sometime after 1.5 million years ago. Why might this be?

From *Homo* to *sapiens*—Talking and Thinking
Lecture 22

It's these early aspects of symbolic behavior that seem so persuasive to us when we think about what it means to be human. They don't approach in scale some of the later evidences of artistic behavior that we have, but they show us that the spark is there. There's something about these populations of people that's different from the humans that came earlier—that's modern in some way.

When we talk about humans, behavioral complexity, and technology, one of the first things that comes to mind is fire. When did humans conquer fire and begin to control it? The earliest evidence we have of human-influenced fire is about 1.4 to 1.5 million years old. We also have evidence, dating from about 1 million years ago on an island called Flores, that humans could cross substantial bodies of water. Did they have boats?

The earliest stone tools were probably not made by people who had language. The larynx of *Australopithecus* was configured very much like the larynx of chimpanzees, which means that there were vocal sacs to help resonate sound, but it is not the kind of larynx that humans have today. However, the hyoid and middle ear bones of early *Homo* show some evidence that it might have had language-like communication.

A gene called FOXP2 helps to establish the pattern of brain development in mammals and may support language in humans. This gene exhibits two differences in modern humans relative to in other mammals, and these differences have also been found in Neandertals.

The Neandertals lived primarily in Europe, about 150,000 to 30,000 years ago. From DNA recovered from two Neandertal individuals, we have learned that Neandertals had one of the O blood types shared by modern humans. They also probably had red hair. Like modern humans, Neandertals from different places were genetically similar. Originally, Neandertals were thought to be a distinct population that emerged in parallel with humans but in a different part of the world. However, recent genetic studies have shown that Neandertal genes are similar enough to the genes of some living people to be able to say that they contributed some of their genes to us.

Interesting differences arise in looking at Neandertals and humans in terms of lifestyles and strategies for subsistence. For example, Neandertal bones are thicker than ours and curved, which means that Neandertals were adapted to more stresses in locomotion than we are. One of the most important aspects of Neandertal biology is the extent to which we find their bones with healed injuries. This suggests that they were not foragers but hunters who used risky and sometimes unsuccessful tactics.

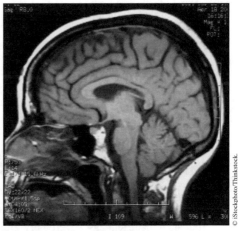

The FOXP2 gene offers clues about when human ancestors developed the brain capacity for language.

A site in Spain shows that Neandertals collected shells, painted them, and strung them from necklaces. They also used chunks of pigment to color on their skin. This kind of symbolic behavior tells us that they were behaviorally very human-like. Similar symbolic behaviors emerged in Africa around the same time.

The fossil remains of humans in Africa after about 200,000 years ago show features that we recognize today in living human populations. The brain had moved forward and the face moved under it. These people had a forehead instead of a brow ridge, and the back of the skull was rounded instead of angled. Such changes may relate to the development of the brain.

By 90,000 years ago, modern humans emerged out of Africa into the Near East, where they may have interacted and exchanged genes with Neandertals. After 60,000 years ago, humans of African origin began to disperse throughout the world. They initially spread across South Asia and, later, into Europe.

We don't really know why humans were ultimately successful and Neandertals weren't. Perhaps the answer lies in their technology, their strategies for hunting, their ability to use resources, or in something we can't see in the archaeological record, such as disease. Such questions present topics for future exploration. ■

Suggested Reading

Guthrie, *The Nature of Paleolithic Art*.

Klein, *The Dawn of Human Culture*.

Questions to Consider

1. In what ways was the disappearance of the Neandertals like historical cases where human populations have come into contact? Are there ways that such historical cases were very different?

2. Language is unique to humans, and it is highly specialized. Can you imagine what precursors to today's human languages may have been like? What features of today's languages might they have had?

Our Accelerating Evolution
Lecture 23

That's a pace of genetic change that's unprecedented in our evolution. If we look at the differences between humans and chimpanzees in years, that's about 10 million years. ... But if we look in terms of mutations that have been selected that have some advantage in populations, the number that occurred between humans and chimpanzees is relatively small compared to the big number that happens in the last 40,000 years.

In this lecture, we look at the ways in which our species, *Homo sapiens*, has changed our society and ecology in the last 10,000 years. These changes are primarily associated with the invention of agriculture and the resulting massive increase in human population. Among the consequences of those developments has been the relatively rapid evolution of many characteristics within our species.

The glaciers of the last Ice Age began to recede about 18,000 years ago, and temperatures began to reach their current levels about 10,000 years ago. Since that time, human populations have grown dramatically, and our species has experienced numerous genetic and physical changes. For example, in the Bronze Age, human skulls had a relatively thick brow ridge, but that shape is rare in living human populations in Europe. Further, over the past 10,000 years in Europe, the size of the brain has shrunk by about 100 cubic centimeters in both males and females and the thickness of the skull has declined.

Genetic changes give us some idea of the ways in which humans adapt to the ecologies they themselves are changing. Consider, for example, the gene that corresponds to the production of lactase, a digestive enzyme that helps us break down lactose, which occurs naturally in milk. In most mammals, this gene is deactivated after weaning because they will never again have access to milk. However, some of our ancestors carried a new copy of the lactase gene that remained active throughout adulthood. This mutation can be traced to populations from the areas where dairy animals became important after the invention of agriculture. Interestingly, we can also tell that the mutations for lactase persistence occurred relatively recently, around 8,000 years ago.

One of the basic factors driving such evolutionary changes is a significant increase in the human population. From the Pleistocene epoch to the present, the world population has grown from perhaps 1 million to 6.5 billion. This population explosion can be traced in various ways in the archaeological record. Evidence of an increase in the tortoise population, for example, tells us that these animals experienced increased predation from humans. Of course, the spread of humans also drove **megafaunal** animals, such as mammoths and mastodons, to extinction.

The rise of cities is another direct sign of ecological change imposed by humans, and this development had a significant influence on human pathogens. Diseases that couldn't get a foothold in a low-density hunter-gatherer population could move as epidemics through humans living in cities. Certain genetic adaptations developed in response to these epidemic diseases, such as those that help us resist particular types of malaria.

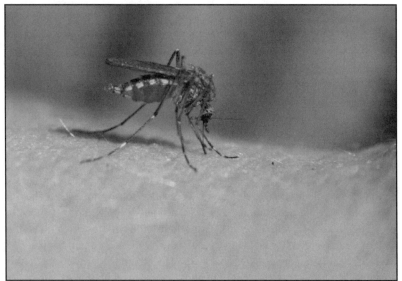

Adaptation can be a response to a disease, such as mosquito-borne malaria.

Hundreds of genes have changed over the last 40,000 years, with such changes intensifying in the last 10,000 years. This pace of genetic change is unprecedented in our evolution. The differences between humans and chimpanzees in years is about 10 million, but the number of mutations that occurred between humans and chimpanzees is relatively small compared to the number that occurred in the last 40,000 years. That small slice of time takes up as much as a quarter to a third of our adaptive evolution. Further, we're imposing rapid selection on many kinds of animals and plants, just as we have created the circumstances for this selection on ourselves. It's a great evolutionary transition in the history of life, one that is centered on what it means to be a modern living human. ■

Important Term

megafaunal: Large animals of a particular region or time.

Suggested Reading

Cochran and Harpending, *The 10,000 Year Explosion*.

McNeill, *Plagues and Peoples*.

Mithen, *After the Ice*.

Questions to Consider

1. A challenge in studying recent human evolution is assessing how human behavior might have been affected by selection. How might our social environments have been changing people during the last 10,000 years?

2. Will culture and technology begin to replace our biological evolution?

Reflections on Major Transitions
Lecture 24

> **What keeps me going ... is that these new fossils provide us with new insights in evolution and how evolution is ... going to evolve.**

T his lecture brings together the perspectives of the two professors for this course—the "deep time" view of the geologist-paleontologist and the paleoanthropologist's approach of working backward from the present.

Among the areas of interest in both fields are the genetic exchanges that must have taken place between early prokaryotes, which gave rise to mutualisms in terms of the development of eukaryotes. At the same time, eukaryotes developed nucleated cells and chromosomes enclosing DNA—genetic defenses that enabled them to resist further encroachment.

One transition not covered in this course was the evolution of grasses and grasslands in the Cretaceous period, which affected primates and numerous other animals. The discovery that dinosaurs were eating grasses at the end of the Cretaceous period was a surprise, because grasses were thought to have originated later, in the Eocene epoch. The increase in grasslands created new habitats for terrestrial primates and restrictions on the habitats of arboreal primates. Ultimately, grasslands were the niche that hominids invaded.

Ecological changes in the early part of the history of life—and today—are massively important. Again, we take grasslands for granted as a biome, but there was a time when they didn't exist. Such ecologies are made by organisms, and when we think of applying what we understand about those processes to the present, the implications are far-reaching.

When we consider evolution in the present and the role of humans in changing evolution, we naturally think about the domestication of plants and animals. Dogs, for example, exhibit tremendous diversity in sizes, colors, shapes, temperaments, and skills, all of which have been selected in the last several hundred years. Some plants and animals that humans have introduced into ecologies have become invasive, spreading rapidly and damaging ecosystems.

Another topic we touched on in the course is coevolution, the development of a mutualism between two organisms. Ants, for example, have domesticated fungus, which they tend in specialized chambers. Ants are also eusocial insects, meaning that they have a differentiation or caste system. It's mind-

Stromatolites are among the most ancient fossils on Earth.

boggling to think about how different castes within a species are selected and how that selection affects other aspects of their lives.

Social organization is, of course, extremely important in primate evolution. Using what we know from the past, what can we say about the future of primate evolution? One study in New England has shown that some traits are different among people who tend to have more children. That's an example of natural selection in ongoing populations in industrialized countries.

> **We should think about human evolution not in terms of stages that we've gone through but as a series of branches of different kinds of related organisms that have each undergone their own history.**

We should think about human evolution not in terms of stages that we've gone through but as a series of branches of different kinds of related organisms that have each undergone their own history. Both paleoanthropology and paleontology apply an interdisciplinary approach to the study of evolution, bringing in genetics and other sciences, to reconstruct those histories. The field of evolution has itself evolved rapidly in the past 40 years and will continue to do so as our understanding of its major transitions deepens. ■

Suggested Reading

Wade (editor), *The New York Times Book of Fossils and Evolution.*

Zimmer, *Evolution.*

Questions to Consider

1. Think about how the evolution of grasslands represents another major transitional event in the history of life. How did this evolution affect the evolution of many organisms, including humans?

2. Evolution is an ongoing process, and will continue as long as life exists on Earth. How could the microevolution observed in organisms today result in their macroevolution, especially with changing environments as a factor?

Life-Cladogram

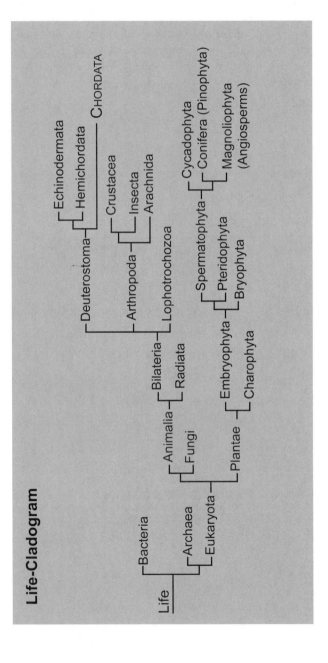

Chordata-Cladogram

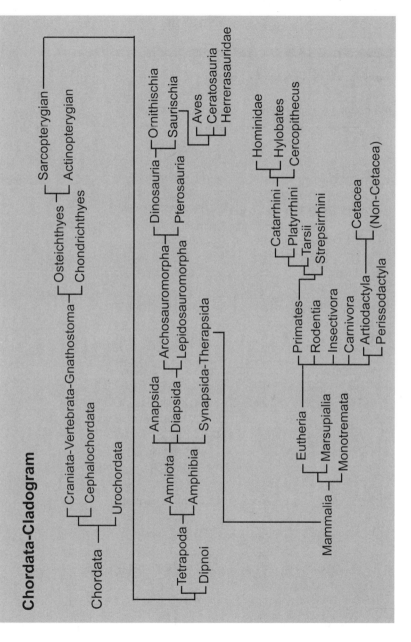

Timeline

Eons, Eras, Periods, and Epochs

4.6 billion years ago (bya) .... Earth forms.

3.8–2.5 bya .... Archean eon.

2.5-0.543 bya .... Proterozoic eon.

0.543 bya–present .... Phanerozoic eon.

543–251 million years ago (mya) ... Paleozoic era.

543–488 mya .... Cambrian period.

488–444 mya .... Ordovician period.

444–416 mya .... Silurian period.

416–359 mya .... Devonian period.

359–318 mya .... Mississippian, or Early Carboniferous, period.

318–299 mya .... Pennsylvanian, or Late Carboniferous, period.

299–251 mya .... Permian period.

251–200 mya .... Triassic period.

200–145 mya .... Jurassic period.

251–65 mya .... Mesozoic era.

145–65 mya.................................... Cretaceous period.

65 mya–present .... Cenozoic era.

65–55 mya..................................... Paleocene epoch.

65–23 mya..................................... Paleogene period.

55–34 mya..................................... Eocene epoch.

34–23 mya..................................... Oligocene epoch.

23–5 mya....................................... Miocene epoch.

23 mya–present .... Neogene period.

5–1.8 mya...................................... Pliocene epoch.

1.8 mya–11,000 years ago............... Pleistocene epoch.

11,000 years ago–present............... Holocene epoch.

Evolutionary Milestones

4.0 bya... Oldest known evidence of life,
from stable carbon isotopes).

3.7 bya... Prokaryote presence, from stable
carbon isotopes (Greenland).

3.5 bya... Photoautotrophs in abundance,
indicated by trace fossils of
stromatolites (Australia).

3.4–3.2 bya.................................... Prokaryote body fossils in chert
(South Africa and Australia).

2.7 bya	Eukaryote presence, from steranes (fossil steroids).
2.1 bya	Fossil algae (probable eukaryotes).
2.0 bya	Molecular clocks predict first metazoans at this time.
1.7 bya	Acritrachs (organic-walled microfossils), resembling some eukaryotes.
1.5 bya	Eukaryote fossils (Australia).
635 mya	Oldest evidence for metazoans from biomarkers (sterols), Canada.
580 mya	Metazoan embryos, China.
580–550 mya	Ediacaran biota, presumed metazoan body fossils and trace fossils from South Australia, Siberia (Russia), Newfoundland (Canada), Namibia, South China, England, and California and North Carolina (USA).
550 mya	*Cloudina*, skeletonized fossil with drillholes in it (probable evidence of predation).
543 mya	Fossil burrow *Treptichnus pedum*, marks the start of the Phanerozoic eon and the beginning of deeper burrowing.
540–530 mya	"Small shelly fossils," a wide range of fossils with uncertain

Timeline

affinities, composed of calcite, silica, and phosphatic minerals.

530 mya.. "Cambrian explosion," the abrupt increase of mineralized fossils in the geologic record, including molluscans, arthropods, echinoderms, and many other animals; *Yunnanozoon lividum*, oldest known chordate (China).

525 mya.. Oldest known trilobites.

513 mya.. Oldest known conodonts (considered vertebrates).

510 mya.. *Anomalocaris*, the largest known predator of the Cambrian period; coincided with trilobite parts in large fecal pellets; *Pikaia*, primitive chordate (Canada).

490 mya.. Oldest known fossil arthropod trackways on land, Cambrian-Ordovician (Canada).

450 mya.. Oldest known fossil arthropod burrows in soils, Ordovician (Pennsylvania).

430–420 mya................................. Molecular clock dates estimated for first insects.

428 mya.. Oldest known arthropod body fossil on land, *Pneumodesmus newmani*, a probable myriapod, Late Silurian (U.K.).

425 mya.. *Cooksonia*, primitive nonvascular land plant in Early Silurian (U.K., Canada, and Australia).

420 mya.. *Baragwanathia longifolia*, oldest known vascular plant.

410 mya.. Fossils of ascomycetes (land fungi), Late Silurian (Sweden); oldest known lycopods (club mosses, quillworts), Late Silurian.

400 mya.. *Aglaophyton major*, nonvascular land plant; *Rhynia gwynne-vaughanii* and *Asteroxylon mackiei*, primitive vascular plants in Early Devonian (Scotland); oldest known insect (and probably winged insect) *Rhyniognatha hirsti*, Early Devonian (Scotland).

395 mya.. Oldest fossil tetrapod tracks, Devonian (Poland).

385 mya.. Lobe-finned fish *Eusthenopteron*, Devonian (Canada).

380 mya.. Lobe-finned fish *Panderichthys*, Devonian (Latvia).

375 mya.. The "fishapod" *Tiktaalik*, Devonian (Canada).

370 mya.. Oldest known pteridosperms ("seed ferns"), Devonian.

365 mya.. Primitive tetrapods *Acanthostega* and *Icthyostega* (Greenland).

359–299 mya.................................. Carboniferous period, the time
of the first true forests.

340 mya... Possibly the oldest known
amniotes, *Casineria kiddi*, Early
Carboniferous (Scotland).

330 mya... Oldest undoubted amniote,
Westlothiana, Late
Carboniferous (Scotland).

325 mya... Previously "oldest" flying insect, Early
Carboniferous (Czech Republic).

315–310 mya.................................. Oldest known amniote fossil tracks,
Late Carboniferous (various locations).

312 mya... *Paleothyris acadiana*, a primitive
amniote, Late Carboniferous
(Nova Scotia); *Hylonomus lyelli*,
another primitive amniote, Late
Carboniferous (Nova Scotia).

310 mya... Oldest known impression (trace fossil)
of a flying insect, Late Carboniferous
(Massachusetts); *Archaeothyris florensis*
and *Clepsydrops collettii*, oldest known
synapsids, Carboniferous (Nova Scotia).

290–272 mya.................................. *Dimetrodon*, various species, large sail-
backed synapsid, Early Permian (USA).

275 mya... *Tetraceratops*, oldest known
therapsid, Early Permian (Texas).

265 mya... Oldest known fossil beetles,
Early Permian (Russia).

250 mya.. *Thrinaxodon*, oldest known cynodonts, Late Permian (South Africa).

245 mya.. Oldest known ichthyosaurs, Early Triassic (various places).

245–237 mya................................... Dinosauromorph tracks, Early-Middle Triassic (various locations).

237–228 mya................................... Bipedal dinosauromorphs, *Lagerpeton* and *Marasuchus*, Middle Triassic (Argentina).

228 mya.. Oldest dinosaur, *Eoraptor lunensis*, Late Triassic (Argentina); Early therapod dinosaur *Herrerasaurus ischigualastensis*, Late Triassic (Argentina).

220(?) mya .... Oldest ornithiscian dinosaur, *Pisanosaurus mertii,* from the Late Triassic of Argentina.

220 mya.. Oldest known fossil flies, Late Triassic (Kyrgyzistan); *Adelobasileus*, oldest known mammal, Late Triassic (Texas).

215 mya.. Theropod dinosaur *Tawa hallae*, Late Triassic (New Mexico).

210–200 mya................................... *Thomasia*, oldest multituberculate mammal, Late Triassic (Germany, France).

210 mya.. Ornithiscian dinosaur *Eocursor parvus*, Late Triassic (South Africa).

200 mya.. *Eudimorphodon*, oldest known pterosaur, Late Triassic (Italy); molecular clocks indicate divergence between gymnosperms and angiosperms, Late Triassic; *Sinoconodon*, primitive mammal, Early Jurassic (China); oldest known plesiosaurs, Late Triassic (U.K., Germany); Gnetales, gymnosperms that share traits with flowering plants (various places).

200–175 mya................................... *Megazostrodon*, primitive mammal, Early Jurassic (South Africa).

170 mya.. Oldest known fossil wasps, Middle Jurassic (Mongolia).

160–155 mya.................................. *Anchiornis huxleyi*, oldest feathered dinosaur, Middle Jurassic (China).

160 mya.. Brachyceran flies that look adapted for flowers, Late Jurassic (China).

150–145 mya.................................. *Archaeopteryx lithographica*, oldest known bird, Late Jurassic (Germany).

130–125 mya.................................. *Eomaia*, oldest known eutherian (placental) mammal, Early Cretaceous (China).

125–120 mya.................................. *Confuciusornis*, oldest beaked bird, Early Cretaceous (China).

125–112 mya.................................. *Repenomamus*, the dinosaur-eating mammal, Early Cretaceous (China).

125 mya.. *Archaefructus*, oldest known
flowering plant, Early Cretaceous
(China); oldest lepidopterans,
Early Cretaceous (Lebanon).

120 mya.. *Kryoryctes*, oldest known
monotreme (Australia).

115 mya.. Oldest freshwater crayfish in
Southern Hemisphere.

110 mya.. *Deinonychus antirrhopus*, theropod
dinosaur with bird-like traits,
Early Cretaceous (USA).

110–105 mya.................................... Oldest known marsupials, various
species, Early Cretaceous (USA).

110–95 mya...................................... Oldest known mosasaurs,
Early-Late Cretaceous.

105 mya.. *Bouliachelys suteri*, oldest known sea
turtle, Early Cretaceous (Australia).

100 mya.. *Melittosphex burmensis*, oldest
bee, Late Cretaceous (Burma).

94–98 mya.. Oldest known snakes, *Eupodophis
descouensi*, Late Cretaceous (Lebanon);
Haasiophis terrasanctus and
Pachyrhachis problematicus (Israel).

60 mya.. Molecular clocks indicate split of first
whales from artiodactyl ancestors.

55 mya.. First true primates.

Timeline

53 mya.. *Pakicetus*, primitive whale, Eocene (Pakistan).

50 mya.. *Ambulocetus*, primitive whale, Eocene (Pakistan).

48 mya.. *Indohyus*, whale-like relative, Eocene (Pakistan).

47 mya.. *Rodhocetus*, primitive whale, Eocene (Pakistan); *Maiacetus*, primitive whale, Eocene (Pakistan); *Darwinius masillae*, primitive primate.

45 mya.. *Eosimias sinensis*, early monkey-like primate.

35 mya.. *Basilosaurus*, primitive whale, Eocene (USA).

32 mya.. *Aegyptopithecus zeuxis*, possible ancestor to apes and Old World monkeys.

26 mya.. Monkeys in South America.

25 mya.. Ape and Old World monkey lineages diverge.

20 mya.. *Proconsul africanus*, an early ape.

12 mya.. *Dryopithecus fontani*, possible ancestor to African apes and humans.

6–4 mya.. Divergence of human and chimpanzee lineages.

4.4 mya.. *Ardipithecus ramidus*, early
hominin ancestor or relative.

4.2 mya.. *Australopithecus anamensis*,
early hominin.

3.6 mya.. Laetoli footprints,
Australopithecus afarensis.

3.2 mya.. Lucy skeleton.

2.8 mya.. Taung skull, *Australopithecus africanus*.

2.6 mya.. First stone tools.

2.5 mya.. Loss of MYH6 functional gene
(jaw muscle reduction).

1.9 mya.. Malapa skeletons,
Australopithecus sediba.

1.8 mya.. Dmanisi fossil skeletons.

1.6 mya.. Nariokotome skeleton.

1.5 mya.. First Acheulean tools.

190,000 ya...................................... First skeletal remains of
modern humans, Ethiopia.

100,000 ya...................................... Early projectile points,
human-altered shells.

60,000 ya.. Ostrich eggshell beads,
personal ornamentation.

50,000 ya.. Modern humans disperse through southern Asia, Australia.

30,000 ya.. Last Neandertals, earliest cave art.

18,000 ya.. Last Glacial Maximum

10,000 ya.. Earliest agriculture, domestication

10,000--5,000 ya .... Falciparum malaria appeared

8,000 ya.. European lactase persistence mutation

6,000 ya.. Neolithic in central Europe

200 ya.. Beginning of Industrial Era

Glossary

absolute age dating: Quantifying the age of geologic materials through radiometric dating or other methods. (See **radioactive isotopes**.)

Acheulean: A stone tool industry dominated by hand axe production, from 1.5 to 300,000 years ago.

acritarch: General term for organic microfossils from the Archean and Proterozoic eons; most are regarded as the remains of single-celled eukaryotes, such as algae.

adaptive radiation: Evolution of a variety of traits (and new species) within a lineage in response to selection pressures, genetic variation, and definition of ecological niches. (See **ecological niche, speciation**.)

allele: One alternate form of a gene.

amniote: A four-limbed vertebrate (tetrapod) that reproduces through internal fertilization and enclosed eggs or live birth; includes reptiles, dinosaurs, birds, and mammals.

amphibian: General term for a four-limbed vertebrate (tetrapod) that reproduces through external fertilization in aquatic or otherwise moist environments.

angiosperm: Flowering plant that reproduces through use of male and female parts in a flower. (See **gymnosperm**.)

anthropoid: Primates, including all apes, monkeys, and humans but excluding prosimians.

archaea: Single-celled prokaryotes; they differ from bacteria through their unique metabolisms and biochemical traits shared with eukaryotes. (See **bacteria**.)

archaeology: Study of past humans through their material artifacts. (See **paleontology**.)

arthropod: Invertebrate animal with jointed legs and chitinous exoskeletons; includes crustaceans, insects, arachnids, and many other groups.

artifact: Something made by people, often tools. The artifactual record is generally associated with the species *Homo*. (See **trace fossil**.)

autotroph: Organism that produces its own food, whether through chemosynthesis or photosynthesis. (See **photoautotroph**.)

bacteria: Single-celled prokaryotes that differ from archaea. (See **archaea**.)

bilateria: Animals with a bilateral symmetry, including all but sponges, comb jellies, and cnidarians (corals, hydrozoans, true jellies).

biological succession: Geological principle discovered in the 19th century by geologists that fossils in vertical sequences of strata show a definite and predictable order, caused by evolution and extinctions. (See **strata**.)

biomarker: Organic compound preserved in the geologic record indicating a biological origin. (See **chemical fossil**.)

biomineralization: The organismal formation of mineralized tissues, such as calcite, aragonite, silica, and apatite.

body fossil: Anatomical remains of an organism in the geologic record. (See **trace fossil** and **chemical fossil**.)

camouflage: Form of visual mimicry, an adaptation used by animals to prevent predation. (See **mimicry**.)

cercopithecoid: An Old World monkey.

chemical fossil: Elements or compounds in the geologic record that are the direct result of an organism. (See **biomarker**.)

chitin: Polysaccharide formed in arthropods and some other organisms for strengthening an exoskeleton.

chloroplasts: Organelles used in photoautotrophs for photosynthesis. (See **photoautotroph**).

choanocyte: Collared cell with flagella in sponges. (See **choanoflagellate**.)

choanoflagellate: Single-celled eukaryote; closely resembles collared cells in sponges. (See **choanocyte**.)

chordate: Animal with a notochord, dorsally located aorta, and pharyngeal gill pouches ("slits"). (See **vertebrate**.)

clade: Group of organisms that share a common ancestor or an organism and all of its descendants. Classifying life in terms of clades (cladistics) is a tool for focusing more directly on shared inheritable traits. Cladistics is now favored by biologists and paleontologists in place of the Linnean classification hierarchy (kingdom, phylum, class, order, family, genus, species). However, in practice (and in this course), the bottom two Linnean taxa (genus, species) are often equivalent to, and themselves used as, clades. (See **cladogram**.)

cladogram: Diagram showing evolutionary relationships between organisms within and between clades; also known as a phylogenetic chart. (See **clade**.)

cleidoic egg: Enclosed egg in amniotes that consists minimally of amnion, allantois, chorion, a yolk sac, and an outer covering (shell).

coevolution: Process in which organisms may exert selection pressures on one another, such as through predator-prey relations or mutualism.

contact: Surface separating two distinctive rock bodies, such as strata.

convergent evolution: Evolution in which similar selection pressures exerted on unrelated organisms have resulted in similar adaptations, e.g., dolphins and ichthyosaurs.

coprolite: Fossilized feces; a type of trace fossil.

developmental biology: Study of how an organism develops during its lifetime; also known as embryology.

diphydonty: Condition in mammals in which only one set of replacement teeth follows an earlier set of teeth.

diploid: Cell containing two sets of chromosomes; typical of somatic (body) cells.

ecological niche: Role of an organism in an ecosystem, which often reflects its adaptations to that ecosystem.

ecology: Study of interactions between non-living materials and organisms.

ectothermy: Physiology in which an organism derives heat from the surrounding environment; sometimes called "cold-blooded."

Ediacaran: Term applied to a Proterozoic fossil assemblage from the Ediacara Hills of South Australia; also used for the last geologic period of the Proterozoic eon.

encephalization quotient: Ratio of the brain to body mass, which provides a way of comparing the brains of animals with different body sizes.

endoskeleton: Internal skeleton, characteristic of all vertebrates and a few invertebrates, such as echinoderms.

endospory: Reproductive process in spore-bearing plants (pteridophytes), in which megaspores (female gametes) are fertilized by microspores (male gametes) within the sporangium of a sporophyte parent; seen as a precursor to the evolution of seed plants (spermatophytes).

endosymbiosis: Hypothesis for the evolution of prokaryotic cells into eukaryotic cells, in which smaller prokaryotes within larger prokaryotes evolved to become organelles and serve other functions within a single cell.

endothermy: Physiology in which an organism generates its own internal body heat; sometimes called "warm-blooded."

entomology: The study of insects, including their evolution and ecology.

environmental change: Global or otherwise large-scale fluctuations in environments through time, such as global cooling or warming.

eon: Largest unit of geologic time, such as the Archean, Proterozoic, and Phanerozoic eons.

epoch: Geologic time unit, subdivision of a period, such as the Paleocene or Eocene epoch.

era: Geologic time unit, subdivision of an eon, such as Paleozoic or Mesozoic.

eukaryote: Organism with a definite cell membrane, genetic material enclosed within a nuclear envelope, and complex organelles; can be either unicellular or multicellular. (See **prokaryote**.)

exaptation: Happenstance of an organism having a favorable trait already in place before a different selection pressure is applied ; sometimes erroneously called a "pre-adaptation."

exoskeleton: External skeleton in which most soft-part anatomy is protected; characteristic of most invertebrates that produce mineralized tissues. (See **endoskeleton**.)

gamete: Reproductive cell (such as sperm or egg), consisting of haploid number of chromosomes .

gene: Sequence of DNA on a chromosome that codes for a trait in an organism.

gene (allele) frequency: The proportion of one allele relative to other alleles in a population.

gene recombination: The exchange of DNA sequences between two copies of a chromosome that may occur when gametes are produced.

genetic drift: Random change in gene frequency, often due to small population size or the founder effect.

genetic map: An account of the locations of genes and DNA sequence markers on chromosomes.

genetics: Study of genes, their changes in populations over time, and differences in genes between different species.

genome: All the genetic material in an organism's complement of chromosomes.

geographic isolation: Separation of different populations of a species by geographic barriers, such as mountains, rivers, or seaways, that prevent them from exchanging genes.

geology: Study of Earth processes and history.

gymnosperm: Seed-bearing plant that does not enclose its seeds in a fruit, thus having "naked seeds." (See **angiosperm**.)

haploid: Cell containing a single set of chromosomes; typical of reproductive cells (gametes).

hemimetabolous: Incomplete metamorphosis in growth (ontogeny) of insects, with major stages of: egg → nymph → adult.

heterospory: Condition in spore-bearing plants (pteridophytes) that have large-sized spores (megaspores) and small-sized spores (microspores); these correspond to female and male gametes, respectively. (See **endospory**.)

heterotroph: Organism that derives its food from external sources, such as through decomposing, herbivory, scavenging, or predation. (See **autotroph**.)

hexapod: Six-legged arthropod, including insects and collembolans (springtails).

holometabolous: Complete metamorphosis in growth (ontogeny) of insects, with major stages of: egg → larva → pupa → adult.

hominin: Clade of primates that includes all species more closely related to humans than to other primates. Includes *Homo* and *Australopithecus*.

hominoid: Primates, including living and fossil apes and humans but excluding monkeys.

homology: A trait that is shared between two clades because they both inherited it from their common ancestor.

Hox gene: An important kind of toolkit gene, shared by all bilaterians, involved in segmenting bodies during embryonic development along the midline axis.

ichnology: Study of organismal traces, such as tracks, trails, burrows, feces, and nests. (See **trace fossil**.)

igneous (rock): Rock formed by crystallization from molten rock.

insect: Six-legged arthropod (hexapod) with a respiratory system and other traits distinct from closely related hexapods, such as springtails.

invertebrate: Animal that does not have a definite mineralized backbone. (See **vertebrate**.)

isotonic: Physiological adaptation of an organism in which it maintains the same solute concentration in its body as water outside its body, helping to cancel out osmotic pressure.

isotope: An alternate form of an element. (See **stable isotopes** or **radioactive isotopes**.)

iterative evolution: Similar trait reappearing in different species within a lineage as a result of similar factors in natural selection and/or the same genes coding for the trait.

lichen: Symbiotic organism comprised of fungi and algae that mutually benefit one another.

macroevolution: Change in gene frequency of a species over time that results in speciation.

mammal: Synapsid characterized by inner ear bones derived from modification of jaw bones in therapsids, as well as mammary glands, hair, and other traits.

marsupial: Mammal that gives live birth but to an underdeveloped embryo, which completes its embryonic growth in a pouch.

mass extinction: A sudden extinction of a significant number of organisms, such as that which occurred at the end of the Permian and Cretaceous periods; marked in the geologic record by an abrupt end to the fossils of many unrelated lineages.

megaspore: Larger spore corresponding to the male gamete in a pteridophyte with heterospory.

meiosis: Cell division from diploid cells into haploid cells, producing gametes.

metamorphic (rock): Rock altered from another rock through heat or pressure.

metazoan: Animal; multicellular eukaryote that can have major changes in its form throughout its ontogeny.

microevolution: Change in the gene frequency of a species over more than one generation.

microspore: Smaller spore corresponding to the female gamete in a pteridophyte with heterospory.

mimicry: Protective adaptation in which one organism imitates another organism through similar appearance, sounds, smells, or other means. (See **camouflage**.)

mitosis: Cell division of haploid cells into identical haploid cells.

molecular clocks: Method that uses rates of molecular change in a lineage and ages of fossil representatives of that lineage to calculate divergence times for two or more related taxa.

molecular phylogeny: Reconstruction of the evolutionary history of a lineage on the basis of molecular (genetic) homology.

monotreme: Modern clade of mammals characterized by egg laying.

mutation: A change to a DNA sequence, which may or may not have functional effects.

mycorrhizae: Fungal symbionts that live around and in the roots of land plants, providing plants with added amounts of phosphorus and nitrogen while gaining carbohydrates from the plant.

natural selection: Charles Darwin's mechanism of evolutionary change, in which organisms survive and reproduce more because they possess particular variations, causing those variations to be more common in the next generation.

notochord: Stiffened structure along the dorsal anatomy of an animal; one of the key traits of a **chordate**.

nucleic acid: DNA (deoxyribonucleic acid) or RNA (ribonucleic acid), which are responsible for transfer and replication of genetic material.

Oldowan: The first stone tool industry, based on simple cores and flakes, beginning around 2.6 million years ago.

ontogeny: Growth history of an organism, which sometimes reflects ancestral traits and, hence, evolutionary history.

osmoregulation: Balance of solutes maintained in an organism relative to solutes outside of its body. (See **isotonic**, **osmosis**.)

osmosis: Movement of water through a semipermeable membrane from an area of greater concentration to one of lesser concentration. (See **isotonic**, **osmoregulation**.)

ovule: Female part of a plant.

paleoenvironment: The ancient environment represented by a paleontological or archaeological site, as reconstructed by faunal or floral remains, ancient soil structure, or water flow characteristics.

paleogeography: Study of ancient landscapes and configurations of continents and oceans; often expressed through paleogeographic maps.

paleontology: Study of ancient life, with "ancient" generally referring to more than 10,000 years ago. (See **archaeology** .)

period: Geologic time unit; subdivision of era.

pharyngeal gill pouches ("slits"): Anatomical feature present in all chordates.

photoautotroph: An organism that produces its own food through photosynthesis.

plate tectonics: Geologic process in which cool, rigid lithospheric plates move above a hot, plastically flowing asthenosphere; results in continental movement, formation and destruction of oceanic crust, mountain building, and volcanism.

pollen: Reproductive part of gymnosperm or angiosperm containing sperm.

pollination: Process by which pollen is introduced to the female part of a plant, often facilitated by animal surrogates, such as insects, birds, or bats.

Precambrian: General term applied to Earth history before the start of the Cambrian period, or about 4.5 billion to 545 million years ago.

primate: A member of the order of mammals that includes apes, monkeys, lemurs, tarsiers, and humans.

prokaryote: Single-celled organism characterized by small size, dispersed genetic material (no nuclear envelope), and no complex organelles. (See **eukaryote**.)

prosimian: A non-anthropoid primate; includes two distinct clades: the strepsirrhine primates and tarsiers.

pteridophyte: Spore-bearing land plant, such as ferns and their relatives. (See **sporangium**, **spore**.)

radiata: Animals with a radial symmetry, such as sponges, comb jellies, true jellies, corals, and hydrozoans. (See **bilateria**.)

radioactive isotopes: Unstable elements that undergo radioactive decay; known rates of decay and ratios of these ("parent element") to a stable end member ("daughter element") can be then used to calculate ages of geologic materials. (See **absolute age dating**.)

Red Queen hypothesis: Hypothesis in evolution that predators and prey coevolve in response to selection pressures exerted by one another, resulting in an evolutionary "arms race"; inspired by a character in Lewis Carroll's *Alice's Adventures in Wonderland*. (See **coevolution**.)

relative age dating: Methods for determining relative ages of rocks; most readily done through application of biological succession. (See **absolute age dating**.)

sedimentary (rock): Rock formed through the cementation of previously eroded rocks or precipitation from an aqueous solution.

seed: Enclosed reproductive part of plant that consists of an embryo, seed coat, and nutrients. (See **spore**.)

snowball Earth: Slang term applied to a continuous span of nearly 100 million years during the Proterozoic eon in which global temperatures were unusually low; this condition probably affected the early evolution of metazoans and other organisms.

speciation: Descent of one or more new species from an ancestral species; can result in adaptive radiation.

sporangia (sporangium): Spore-bearing structures in a fungus or plant.

spore: Reproductive structure in a fungus or plant that disperses easily (through wind or water); has very little to no stored food for the reproductive cell. (See **pteridophyte**, **seed**.)

stable isotopes: Elements that do not undergo radioactive decay but have different atomic masses, such as carbon-12 and carbon-13 or oxygen-16 and oxygen-18. Used as chemical fossils to reconstruct diet or paleoenvironment.

stoma (stomata): Pores surrounded by specialized cells in the leaves of land plants; they regulate respiration and prevent water loss.

strata (stratum): Layers of sedimentary rock.

stromatolite: Layered sedimentary structure formed by trapping and binding of sediment in a bacterial or algal colony; trace fossil.

tetrapod: Four-limbed vertebrate, descended from lobe-finned fish.

therapsid: Synapsid reptiles that lived from the Permian through the Triassic periods; considered to be immediate ancestors of mammals because of their mammal-like traits.

trace fossil: Indirect evidence of ancient life formed as a result of behavior, such as tracks, burrows, tooth marks, or feces. (See **body fossil**, **ichnology**.)

transitional fossil: Popular term applied to a body fossil that shows a blend of traits between two lineages, such as *Archaeopteryx*, with its dinosaurian and avian traits. The term is considered unnecessary by most evolutionary scientists because every fossil represents a transition between generations.

uniformitarianism: Principle in geology, first defined by Charles Lyell in the 19th century, that most Earth processes and their products observed today are comparable to those of the geologic past.

vertebrate: Chordate that has additional mineralized tissue supporting its notochord ("backbone").

zooanthellae: Symbiotic algae that live within tissues of corals; responsible for giving corals oxygen needed for their growth while corals provide the algae with living space.

Bibliography

Ahlberg, P. E., ed. *Major Events in Early Vertebrate Evolution: Palaeontology, Phylogeny, Genetics, and Development*. Boca Raton, FL: CRC Press, 2001. In 23 chapters, a wide variety of researchers from different realms of evolutionary science—such as paleontology, genetics, and developmental biology—lend their perspectives to unraveling milestones in early vertebrate evolution, beginning with the transition of invertebrates to chordates and from lobe-finned fish to tetrapods.

Beard, Christopher. *The Hunt for the Dawn Monkey: Unearthing the Origins of Monkeys, Apes, and Humans*. Berkeley: University of California Press, 2006. Beard relates the research on early primates from Cuvier to his own discovery of Eosimias, the oldest known anthropoid primate. A good description of the ways in which field paleontology progressed in different parts of the world.

Beerling, D. *Emerald Planet: How Plants Changed Earth's History*. Oxford: Oxford University Press, 2008. The author takes the viewpoint that land plants have not only evolved in response to changing environments but have also played a role in altering environments over the course of geologic time. Similar to Kendrick and Davis's book, this one covers the evolution of plants in more or less chronological order but looks more at major events in Earth history related to plants.

Benton, M. J. *Vertebrate Palaeontology* (3rd ed.). New York: Wiley-Blackwell, 2005. This book is meant as an advanced undergraduate text for paleontology students interested in the fossil record of vertebrates, but the first chapter also serves as an effective summary of current hypotheses on the origin of vertebrates from invertebrates.

Berta, A., J. L. Sumich, and K. M. Kovacs. *Marine Mammals: Evolutionary Biology*. New York: Academic Press, 2006. Although this book includes marine mammals other than cetaceans, one chapter nonetheless gives a concise assessment of the evolution of cetaceans. Explanations of more general concepts related to the anatomical and behavioral adaptations of sea-dwelling mammals assist in understanding how they could have evolved from land-dwelling mammals.

Carpenter, K. *Dinosaur Eggs, Nests, and Babies*. Bloomington: Indiana University Press, 1999. This book is primarily about the fossil evidence related to dinosaur reproduction, but it also contains a good deal of information about amniote eggs in general, including a chapter that considers the origin of the enclosed amniotic egg.

Chiappe, L. M. *Glorified Dinosaurs: The Origin and Early Evolution of Birds*. New York: Wiley-Blackwell, 2005. Chiappe, one of the top researchers on fossil birds from the Mesozoic era, provides a concise, well-illustrated, and up-to-date assessment of the fossil evidence supporting the evolutionary transition of birds from theropod dinosaurs and their subsequent evolutionary history.

Clack, J. A. *Gaining Ground: The Origin and Evolution of Tetrapods*. Bloomington: Indiana University Press, 2002. Clack has spent most of her career researching the origin and evolution of early tetrapods from lobe-finned fish. Thus, her coverage of this topic stems from a personal perspective but is filled with her knowledge and experience in the study of Devonian tetrapod fossils and how these relate to the evolution of four-limbed vertebrates on land.

Cochran, Gregory, and Henry Harpending. *The 10,000 Year Explosion: How Civilization Accelerated Human Evolution*. New York: Basic Books, 2009. Cochran and Harpending explore the genetics of recent human evolution, documenting the rapid natural selection following the invention of agriculture.

Dudley, R. *The Biomechanics of Insect Flight: Form, Function, Evolution*. Princeton: Princeton University Press, 2002. Dudley provides a broad but detailed look at the major selection factors associated with the evolution of insect flight, such as non-winged ancestors, aerodynamics, energetics, diversification, and pollination. He even considers the future of insect flight, an example of how evolutionary theory can be applied to predict upcoming trends.

Everhart, M. J. *Oceans of Kansas: A Natural History of the Western Interior Sea*. Bloomington: Indiana University Press, 2005. Filled with personal insights gained from decades of field work and study of fossil specimens, Everhart's book enthusiastically paints a picture of marine life during the late Cretaceous in the area of what we now call Kansas, with emphasis on the marine and flying reptiles of that region and time.

Fedonkin, M., J. G. Gehling, K. Grey, G. M. Narbonne, and P. Vickers-Rich. *The Rise of Animals: Evolution and Diversification of the Kingdom Animalia*. Baltimore, MD: Johns Hopkins University Press, 2007. Up-to-date coverage of the latest research on the origin of Precambrian animals and other life, written with lively text and well illustrated. These experienced and well-traveled authors summarize the fossil record for Ediacaran animals throughout the world, while also supplying a primer on the first 4 billion years of Earth history.

Fraser, N., and D. Henderson. *Dawn of the Dinosaurs: Life in the Triassic*. Bloomington: Indiana University Press, 2009. This book is about more than just the early evolution of dinosaurs; it also discusses many of the plants, other animals, and environments throughout the Triassic world. Fraser and Henderson work collaboratively as scientist and artist, respectively, to re-create Triassic landscapes that share recognizable elements with modern ones, while setting the stage for the remainder of the Mesozoic era, also known as the "age of dinosaurs."

Gensel, P. G., and D. Edwards, eds. *Plants Invade the Land: Evolutionary and Environmental Perspectives*. New York: Columbia University Press, 2001. Twenty-six paleontologists and geologists cover, in 13 chapters, the probable evolutionary pathways and ecological consequences of plants adapting to terrestrial environments from the Ordovician through the Devonian periods.

Gibbons, Ann. *The First Human: The Race to Discover Our Earliest Ancestors*. Garden City, NY: Anchor, 2008. An able reporter, Gibbons goes on scene with the teams finding the earliest known hominins. Her story shows the excitement accompanying the announcements of the fossils and the conflicts between these teams as they argue about their interpretations. A great portrait of the science of early human evolution.

Gould, S. J. *Wonderful Life: The Burgess Shale and the Nature of History*. New York: W.W. Norton, 1989. Through its description and commentary on the Burgess Shale fauna, this book introduces us to what animals looked like before skeletons became more common after the Cambrian period. Although the book is now outdated in some respects and some of its factual information has since been corrected, Gould, using his engaging and informed prose, pulls in readers to marvel with him at the amazing biological diversity represented by this fossil assemblage.

Grimaldi, D. A., and M. S. Engel. *Evolution of the Insects*. Cambridge: Cambridge University Press, 2005. A comprehensive volume on the fossil record of insects by two paleoentolomogists who made many of the primary discoveries discussed here. The text is augmented by many illustrations of fossil insects, including photographs of exquisitely preserved insects in amber from the Mesozoic and Cenozoic eras. The book also deals with the origin and evolution of the various insect groups associated with pollination during the later part of the Mesozoic era, such as hymenopterans (wasps, bees, and ants), coleopterans (beetles), dipterans (flies), and lepidopterans (moths and butterflies).

Guthrie, R. Dale. *The Nature of Paleolithic Art*. Chicago: University of Chicago Press, 2006. A favorite of one of the professors. Guthrie is an archaeologist who has spent years getting into the minds of ancient hunters—replicating their techniques and thinking about their relationships with animals. He has used this knowledge to interpret ancient art. The book is richly illustrated by Guthrie's own drawings reproducing these ancient artworks.

Johanson, Donald, and Kate Wong. *Lucy's Legacy: The Quest for Human Origins*. New York: Three Rivers Press, 2010. Johanson has been a leader in paleoanthropology for nearly 40 years. In this book, he describes the important fossil discoveries of the last 15 years, from his team's return to Ethiopia to the exciting finds from Dmanisi, Georgia.

Kemp, T. S. *The Origin and Evolution of Mammals*. Oxford: Oxford University Press, 2005. A detailed account of the synapsid-reptile ancestors of mammals, the anatomical and physiological changes that took place in the transition between these reptiles and mammals, and the important clues provided by living mammals about their evolutionary lineage.

Kenrick, P., and P. Davis. *Fossil Plants*. Washington, DC: Smithsonian Books, 2004. A succinct introduction to paleobotany, with chapters arranged in order of geologic time and covering bacteria, fungi, algae, lichens, and land plants. For those who might not have much background in botany, the glossary of this book is especially helpful for understanding material in this course related to the macroevolution of land plants.

Kielan-Jaworowska, Z., R. F. Cifelli, and Z.-X. Luo. *Mammals from the Age of Dinosaurs: Origins, Evolution, and Structure*. New York: Columbia University Press, 2006. Overview of Mesozoic mammals, including the factors that likely led to their evolution from non-mammalian synapsids in the Triassic period, their subsequent diversification into many groups of mammals, and the locations where their fossils have been found.

Klein, Richard G. *The Dawn of Human Culture*. New York: Wiley, 2002. Klein specializes in the archaeology of African sites from around the time of the origin of modern human populations. He explores the question: What made these Africans succeed and spread throughout the world?

Levin, H. *The Earth through Time* (9th ed.). New York: Wiley-Blackwell, 2009. This textbook, intended for undergraduate students, provides an introduction to historical geology, including an overview of plate tectonics, relative age dating, absolute age dating, and the fossil record. Evolution and extinctions are placed in the context of shifting continents and fluctuating sea levels, giving a large-scale perspective on the occurrence of natural selection in the geologic past.

Levinton, J. S. *Genetics, Paleontology, and Macroevolution*. Cambridge: Cambridge University Press, 2001. This book shows how genetics, paleontology, and developmental biology can be integrated to better discern the patterns and processes of macroevolution. Levinton also notes that many of the original concepts outlined by Darwin (adaptations, variations in form, and selection in populations) can be explored further through this updated information.

Mann, S. *Biomineralization: Principles and Concepts in Bioinorganic Materials Chemistry.* Oxford: Oxford University Press, 2001. This book is meant more for advanced undergraduate or graduate education, rather than a general audience. Nevertheless, it effectively explains the chemistry and biochemistry of the secretion of minerals in organisms' exoskeletons or endoskeletons, the types of minerals formed, and the relationship of biomineralization to evolutionary processes.

Margulis, L., and M. Dolan. *Early Life: Evolution on the Precambrian Earth* (2nd ed.). Sudbury, MA: Jones and Bartlett Learning, 2002. The primary author of this book (Margulis) originated the endosymbiotic hypothesis as an explanation for the evolution of eukaryotes from prokaryotes; hence, she is a most appropriate authority for an examination of Precambrian microbial evolution. This updated edition of the 1984 book includes new information relevant to the evolution of single-celled organisms.

Martin, A. J. *Introduction to the Study of Dinosaurs.* New York: Wiley-Blackwell, 2006. This book, intended as a textbook for undergraduate non-science majors, gives an overview of dinosaurs and their evolution throughout the Mesozoic era. It also includes an extensive chapter about their early evolution from non-dinosaur reptiles during the Triassic period.

McNeill, William H. *Plagues and Peoples.* Garden City, NY: Anchor, 1977. McNeill's book is a true classic. Many of the ideas have been updated in the intervening years, but nobody has done a better job of putting together the story of disease and its effects on history.

Mithen, Steven. *After the Ice: A Global Human History, 20,000–5000 BC.* Cambridge, MA: Harvard University Press, 2006. Mithen gives an account of the archaeology that shows hunter-gatherers in the wake of the last glaciation, at the dawn of the development of agriculture. He weaves different parts of the world together into a story about the changes in societies with this new lifestyle.

Morris, S. C. *The Crucible of Creation: The Burgess Shale and the Rise of Animals.* Oxford: Oxford University Press, 1999. Morris, a foremost researcher on the Burgess Shale fauna, takes us several steps further than Gould with his assessment of this fossil assemblage and the insights it provides into the evolution of animals before, during, and after the Cambrian period. Morris also adds information about other Cambrian fossil assemblages comparable to those of the Burgess Shale, while critiquing some of Gould's views on the latter.

Prothero, D. R. *Evolution: What the Fossils Say and Why It Matters*. New York: Columbia University Press, 2007. Prothero provides a comprehensive look at examples of fossils that best illustrate evolutionary transitions in animals. Through well-organized, clear explanations of the evidence for macroevolution, as well as his inclusion of numerous images of transitional fossils, he also directly addresses and refutes most of the standard arguments leveled against macroevolutionary theory.

————, and S. E. Foss, eds. *The Evolution of Artiodactyls*. Baltimore, MD: Johns Hopkins University Press, 2007. Various researchers discuss the fossil record for the evolutionarily related group of hoofed mammals known as artiodactyls, including whales. Specifically, the authors of the third chapter (Jonathon Geisler and others) explain the evolutionary connections between land-dwelling artiodactyls and whales, as indicated by shared traits in the comparative anatomy of fossil whales and modern artiodactyls.

Rizzotti, M. *Early Evolution: From the Appearance of the First Cell to the First Modern Organisms*. Basel, Switzerland: Birkhäuser, 2000. Rizzotti, through his review of the traits of prokaryotic and eukaryotic cells, asks important questions about the evolution of single-celled organisms and how their evolution eventually led to the evolution of multicelled organisms.

Shear, W. M. "The Early Development of Terrestrial Ecosystems." In H. Gee, ed., *Shaking the Tree: Readings from Nature in the History of Life*, pp. 169–183. Chicago: University of Chicago Press, 2000. An overview of the evolutionary factors that made possible the transition of life from aquatic to terrestrial environments in the first half of the Paleozoic era and the life forms that were among the earliest colonizers of these ecosystems.

Shipman, P. *Taking Wing:* Archaeopteryx *and the Evolution of Bird Flight.* New York: Simon & Schuster, 1999. This book is a little outdated in some of its content, considering how many transitional fossils of feathered dinosaurs and Mesozoic birds have been found since 1999. Nonetheless, it is still worth reading for Shipman's clearly written and well-researched summary of what was known about *Archaeopteryx*—the first known bird from the late Jurassic—and how paleontologists have studied its connection to the evolution of flight in this lineage of what we now see as "modern dinosaurs."

Shubin, N. *Your Inner Fish: A Journey into the 3.5 Billion–Year History of the Human Body.* New York: Pantheon, 2008. In what is arguably one of the best paleontology books written for a general audience in recent years, Shubin explores, through easily understood explanations, concepts, and analogies, the anatomical and genetic evidence for the evolution of our species from lobe-finned fish and their tetrapod descendants.

Stanley, S. M. *Earth System History* (3rd ed.). New York: W.H. Freeman, 2008. Stanley's textbook is intended for undergraduate historical geology classes, but it emphasizes the relationship of biological evolution to changes in ecological and geological systems through time. One of the most important concepts put forth in this book is the idea that some organisms also altered their environments, thus affecting the course of evolution in unrelated lineages.

Sumida, S. S., and K. L. M. Martin, eds. *Amniote Origins: Completing the Transition to Land.* New York: Academic Press, 1997. Although this volume is now dated and has been succeeded by many important fossil discoveries since its publication, its 24 authors cover the most important concepts related to the transition of water-dwelling tetrapods to fully terrestrial lifestyles. Physiology, behavior, anatomy, biogeography, ecological change through time, and other factors are connected to the fossil record of early amniotes.

Taylor, Jeremy. *Not a Chimp: The Hunt to Find the Genes That Make Us Human.* New York: Oxford University Press, 2010. Taylor, a BBC science editor, explores the science of genetics and psychology to find what we think makes humans different from chimpanzees. The book puts behavior into the context of biology.

Taylor, T. N., E. L. Taylor, and M. Krings. *Paleobotany: The Biology and Evolution of Fossil Plants* (2nd ed.). New York: Academic Press, New York, 2009. At 1,230 pages, this book is obviously a much more ambitious coverage of paleobotany than the previously recommended book by Kenrick and Davis. However, readers can focus on the excellent, comprehensive section dealing with the early evolution of flowering plants, starting at page 873.

Unwin, D. *The Pterosaurs: From Deep Time*. Harlow, UK: Pi Press, 2006. Unwin is one of the world's leading researchers of pterosaurs—flying reptiles of the Mesozoic era. Hence, few authors could cover this topic as thoroughly yet without overwhelming a reader interested in the evolution of the first flying vertebrates. The illustrations (many in color), which bring these animals back to life in the reader's imagination, are an added bonus to the text.

Vickers-Rich, P., and P. Komarower, eds. *The Rise and Fall of the Ediacaran Biota*. Special Publication 268. London: Geological Society of London, 2009. This book is a compilation of articles by the top researchers on Precambrian metazoans and their geological context, giving readers the latest hypotheses concerning Proterozoic climate, plate tectonics, and enigmatic Ediacaran fossils. Although intended more for specialists in Precambrian paleontology, it is a good follow-up for those wanting more details after reading *The Rise of Animals* (Fedonkin, et al.).

Wade, N., ed. *The New York Times Book of Fossils and Evolution*. New York: Globe Pequot, 2001. An anthology of news stories published in *The New York Times* "Science Times" section that largely dealt with then-newly discovered evolutionary transitions indicated by the fossil record, from Precambrian animals to humans. Readers may enjoy comparing what were "the latest" discoveries as of 2000 to what we now know—through this course and current events—effectively showing how paleontology, like all sciences, also "evolves."

Walker, Alan, and Pat Shipman. *The Ape in the Tree: An Intellectual and Natural History of Proconsul*. Cambridge, MA: Belknap Press of Harvard University Press, 2005. The book's title refers to an exceptional find of a fossil ape inside a fossil tree. Walker and Shipman give a lively history of the personalities who dug up the fossil evidence of Miocene apes.

Zimmer, C. *Evolution: The Triumph of an Idea*. New York: Harper Collins, 2002. Zimmer, one of the most well regarded popular science writers, gives an excellent overview of the history and concepts of evolutionary theory through his lucid writing and illustrations. This book was written to accompany the eight-part PBS series with the same name and augments the fossil record with other major sources of evidence for evolutionary theory.

————. *Smithsonian Intimate Guide to Human Origins*. New York: Harper Paperbacks, 2007. Zimmer's book is a well-illustrated, broad overview of human evolution, covering the whole time span.

Web Sites

American Geological Institute: Evolution and the Fossil Record. http://www.agiweb.org/news/evolution/datingfossilrecord.html.

University of California Museum of Paleontology. http://www.ucmp.berkeley.edu/.

University of California Museum of Paleontology: Understanding Evolution. http://evolution.berkeley.edu/.

The Virtual Fossil Museum: Fossils across Geological Time and Evolution.

Notes

Notes